MW00736840

IN THE WATERSHED
A Journey Down the Maumee River

Ryan Schnurr

Copyright © 2017 Ryan Schnurr

All rights reserved. This book or any portion thereof may
not be reproduced or used in any manner whatsoever
without the express written permission of the publisher
except for the use of brief quotations in a book review.

First edition 2017

ISBN: 978-0-9989041-0-8

Belt Publishing
1667 E. 40th Street #1G1
Cleveland, Ohio 44120
www.beltmag.com

Book design by Meredith Pangrace
Cover by David Wilson

For Anna

how the water goes
is how the earth is shaped.

—*Jim Harrison*

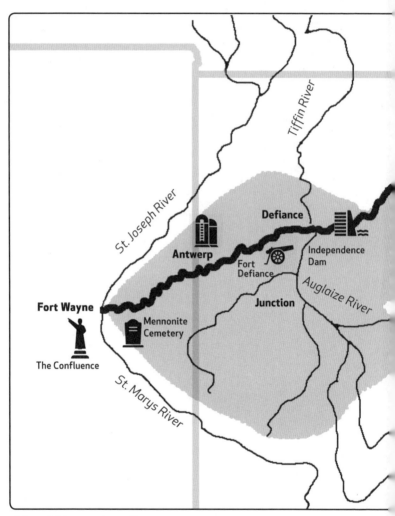

Map of Maumee River Journey

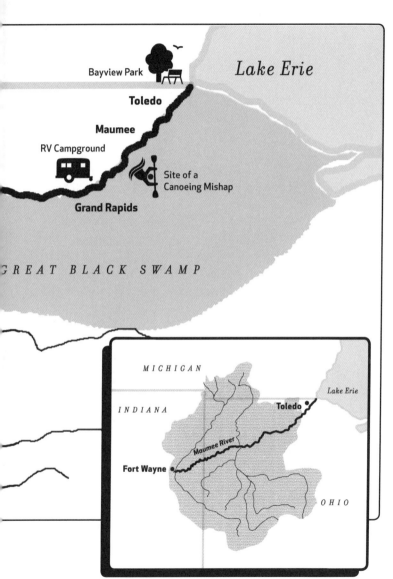

Bayview Park

Lake Erie

Toledo

Maumee

RV Campground

Site of a
Canoeing Mishap

Grand Rapids

GREAT BLACK SWAMP

MICHIGAN

INDIANA

Lake Erie

Toledo

Maumee River

Fort Wayne

OHIO

Overview of Maumee Watershed

CHAPTER 1

The Maumee River does not begin. Formed out of the confluence of the St. Joseph River from the north, and the St. Marys River from the south, it is a continuation, flowing eastward and slightly northward through northeast Indiana and northwest Ohio, eventually opening out 137 miles later into the southwest corner of Lake Erie. Find the river on a map, and it looks to be divided into two distinct parts. The first, from its headwaters in Fort Wayne, Indiana, to its midpoint near Defiance, Ohio, is narrow, between one hundred and two hundred feet wide, and unfolds itself in dozens of meandering curves coiled tightly, one after another, like a stretched-out spring. The second, from midpoint to mouth, is wider and straighter. If your map is topographical, you will see that the whole thing sits in the middle of a broad, rather flat area of land that stretches out from the river in all directions. What you are looking at is the Maumee River Basin, the largest watershed in the Great Lakes region, collecting runoff from more than sixty-six hundred square miles of land in Indiana, Ohio, and Michigan, and sending it downstream.

Among the tributaries that feed into the Maumee River are the Auglaize River, the Blanchard River, the Tiffin River, the Ottawa River, and of course, the St. Joseph and the St. Marys. But these are only the biggest. The larger tributaries are in turn fed by smaller ones, dozens of them, with names like Swan Creek, Powell Creek, Flatrock Creek, Beaver Creek, Bad Creek, Tontogany Creek, and South Turkeyfoot Creek. Each stream drains its own, smaller watershed, adding up to the Maumee River Basin. The whole system—four thousand miles of it—stretches out to network the region like veins on a leaf. These particular veins are the recipients of an estimated four hundred billion gallons of runoff each year, most of it during heavy spring rains. The runoff brings with it wastewater, topsoil, and the fertilizers and pesticides that people living in the watershed dump onto the ground, not to mention whatever is dumped into the streams directly. Most of this eventually ends up in the Maumee River, which is responsible for more than 50 percent of the sediment, and a considerable amount of the nitrogen and phosphorous, that makes its way into Lake Erie, a body of water not known for its cleanliness.

In the middle part of the 1990s, massive, toxic algae blooms of increasing severity began appearing on the shores of Lake Erie, forming giant "dead zones," which, as the name suggests, cannot sustain life—huge swaths of the nation's largest concentration of freshwater turned lethal. The algae blooms are fed by high levels of, among other things, nitrogen and phosphorous.

My interest in this matter is not purely intellectual. I grew up in Fort Wayne, Indiana, a city that happens to be located at the confluence of the three rivers. Downtown lies along the south bank of the St. Marys, its northeast corner butting up against the headwaters of the Maumee. A curious thing about living in Fort Wayne is that if it weren't for the fact that every third business begins with the words "Three Rivers," you would never guess that this was the case. It's quite difficult to see any of the rivers, save for a few scattered footbridges. It is entirely possible that a person could visit the city, drive along the St. Joseph, park near the Maumee, and eat dinner next to the St. Marys, without noticing any of them.

For a long time, the rivers were seen as threatening by Fort Wayne's residents, especially as neighborhoods were built right up next to them in the flood plains. In 1913, the Maumee overflowed into the homes and businesses on its north shore. In 1982, another major flood interrupted neighborhoods along the St. Marys. The solution was to construct a series of twenty-five-foot earthen levees along the banks. Beyond concerns over flooding were those of pollution. The St. Joseph River supplies a good portion of the water for the municipality, and the barriers keep residents from thinking about what the water looks like. The rivers are not particularly attractive—they are a sort of dull, milky brown—and popular wisdom held that they concealed "three-eyed fish" and various other kinds of uncanny creatures. "Don't go in the river," adults half-jokingly warned teenagers when I was growing up. "You might come out with a third arm." The stretches of waterfront that aren't behind levees are, with a few exceptions, obscured by trees and buildings and concrete walls.

In the fall of 2013, my wife, Anna, and I moved near downtown Fort Wayne, into the second floor of a medium-grey Queen Anne on Columbia Avenue. The house sat on the northern banks of the Maumee, less than a quarter mile from its headwaters. From the front windows of the apartment I could see down a narrow side street to a twenty-foot embankment. On top of this thick earthen wall was a stretch of cement path, a section of the city's Rivergreenway trail system. On the other side was the river. I was bike commuting at the time, and would often ride these trails. The trails are one of the only ways to get a good look at the Maumee, and I began to wonder about it. The writer Robert Sayre claims that most people know astonishingly little about the rivers around them, their histories, watersheds, pollution, and sources of pollution. I was no exception. In fact, I knew more about Lake Michigan and the Mississippi than I did about the river that passed just one hundred yards or so from my living room.

I began to read widely on the topic: early geographies, local histories, news coverage, EPA reports. I emailed river experts and advocates, and attended a meeting of the organization Save Maumee, where I drank a free glass of iced tea, read four different brochures, and listened to a discussion about invasive species removal. At the downtown branch of the Allen County Public Library, I found an entire row of shelves dedicated to local histories for northeast Indiana and northwest Ohio that included documents like land deeds and court cases. I pulled twenty-five books off the shelf, made a rough mental estimate of the number of pages in those twenty-five books, and put fifteen back. I read one book on Little Turtle, a Miami war chief; waded slowly through *History of the Maumee River Basin*, a 634-page behemoth written in 1905; and flipped through a legal petition, dated January, 1852, by the descendants of the first miller to set up shop in the area, who were asking the House of Representatives for damages related to the destruction of their grandfather's mill. (Their petition was denied.)

The river was seeping into my everyday life. I started to dream about it, and it surfaced more and more often in casual conversation, and at dinner with friends. (This development had its negative aspects; in my experience, you can only say "riparian buffer" so many times in a conversation before the other party excuses herself.) I was learning a great deal about the Maumee, its history, watershed, pollution, and sources of pollution. But there was one thing glaringly absent from my research: the river itself. You can learn a lot from books, and also from experience. One is not always a very good substitute for the other, and it was clear that my knowledge of the Maumee was still limited, still abstract. I began to have a recurring thought—a curiosity, really—which devolved into an undertaking: what if I walked the river from one end to the other?

In late summer of 2016, I did.

———————

Officially, rivers fall under the category of streams—"linear flowing bodies of water." Unofficially, a river is a whole lot more.

Rivers run through the center of human and nonhuman life. People have historically tended to organize themselves around water, for reasons, among others, of transportation and hydration. (Rivers being excellent for each.) On top of that, the whole appearance of any given landscape is mainly due to the influence of water in one form or another. Glaciers, rivers, rain—these have spent the better part of history clearing prairies, cutting valleys, sculpting the country into what it looks like today. It works the other way, too—a river shapes but it is also shaped. Rainfall, temperature, erosion, water diversion, new tributaries, and construction like dams or canals: all can change the course and dimensions of a river. Change can happen over a long time—decades of climate change, say—or all of a sudden; anyone who's lived through a flash flood can tell you how. Everything that happens in a watershed happens to its river.

One of the first things I read about the Maumee River was that it was named for the Miami people, who had their largest village, Kekionga, at the headwaters. Maumee is an English transcription of the Ottawa word for the Miami people, *(o) maamii*. The Miami called the river *taawaawa siipiiwi*, after the Ottawa. The French, once they arrived, called the river *la riviére aux-Mis*, which translates to "the River of the Miami." In some early treaties, the Maumee was called the Miami of Lake Erie, or Miami of the Lake. "Kekionga" is also a product of English transcription about which there is some inexactness. For a while, historians thought that it meant something close to "blackberry bush," or "blackberry patch." Some people have believed that the settlement's name was a variation of the word *Kiskakon*, which meant "hair clipping place," or the place where warriors would shave their heads to prepare for battle. The most convincing answer, and the one used by the Miami themselves, is that the original word was *Kiihkayonki*, which meant, roughly, "the place of the elders."

Kiihkayonki sat near the northern banks of the Maumee, just east of its headwaters. This was a strategic site. It gave the Miami access to—and influence over—everything that happened at the crossroads of the three rivers. Before highways or train tracks, rivers dictated primary travel routes, and the Maumee was right in the middle of an important one. Using this route, a group of traders in the eighteenth century could go across the continent, from settlements on the Great Lakes to the Gulf of Mexico, by water: beginning somewhere on Lake Huron, for example, they would head southward into Lake Erie and then up the Maumee to its headwaters. Then they would get out and portage, carrying their bark canoes and bundles of fur eight miles southwest to the Little River, which fed into the Wabash. The Wabash would take them west and south to the Mississippi, which in turn went all the way to the Gulf. Alternatively, if they were coming from farther east—even as far east as the Atlantic Ocean—they could begin at the mouth of the St. Lawrence River, or Lake Ontario, and follow the same path through Lake Erie and on to the Wabash. But no matter how they got there, they had to pass by the Miami.

At Kiihkayonki, residents lived in domed log structures covered in bark, called wiccias, surrounded by gardens. They raised squash, beans, and melons. They wore clothing made of animal skins. Eventually, they wore European-style cloth. Their skin was lighter than some other tribes. Men cut their hair short, but kept long locks around the ears. Women often had tattoos on their cheeks and chins. They ate a lot of corn. Fields of corn and cattle stretched out from the headwaters in all directions. The area around the confluence had a nearly sacred significance for the Miami. Families would travel there in springtime to work the fields and prepare for war. The Miami war chief *Mihsihkinaahkwa*, known by the English name Little Turtle, called this place "That glorious gate, through which all good words of our chiefs had to pass from the north to the south and from the east to the west."

Now there is a dam at Mihsihkinaahkwa's glorious gate. It is on the St. Joseph River, just upstream of the confluence. The dam was built in 1933 in conjunction with a water treatment plant, which distributes tap water to most of the city. The engineers who chose the site first looked at the Maumee, but decided against it, given that most of the city's drainage, including raw sewage, went that way. The engineers knew that it was important to keep as much raw sewage as possible out of a city's water supply.

This does not mean that the water in the St. Joseph River is clean. At the dam, "raw" river water is pumped out and carried through long pipes, forty-two inches in diameter, to the Three Rivers Water Filtration Plant, where it is combined with ferric sulfate, lime, and carbon in a process called flocculation. This plant uses big paddles that beat in the chemicals like a giant hand mixer. The idea behind flocculation is to catch up all the soil particles and other bits of gunk into clumps, called "floc." Ferric sulfate helps this happen. (The lime is to soften the water, and the carbon soaks up pesticides and fertilizers and makes the water smell and taste better.) After flocculation, the water goes into "settling tanks," where the floc sink to the bottom. The water on top is drained off and flocculated again, and then gets a chemical treatment of chlorine and fluoride. If you put a glass of this water next to a glass of water dipped straight out of the river, you would probably wonder what in the world people drank before the Three Rivers Water Filtration Plant.

The short answer is that, before the 1870s, people in Fort Wayne got their water from the rivers. The long answer is that they also got it from springs, creeks, wells, and cisterns. But by the middle of the nineteenth century, that system was no longer working. There were getting to be too many people, which caused problems of the water-quality variety. A main reason for this was that most of the toilets were outside, close to the water. The water that people were drinking started to carry diseases like typhoid and cholera. By 1875, residents of the city were clamoring for a proper water system to be built. The owners of an old feeder canal offered to build their own system and sell water to the city. The city council was in favor of this idea, but most of the citizens were not. As it happened, 1875 was an election year, and the people voted out every candidate who favored the plan to buy water from the canal.

The incoming council hired a hydraulic engineer to put together a new plan for treating the water. It included a steam-powered pump on a tributary to the St. Joseph called Spy Run Creek. But they had miscalculated how much water they would be able to get out of the creek. A drought dried it up, and Fort Wayne had to buy water from the owners of the old feeder canal anyway. Later, the city drilled deep wells into a water table under the town. Fort Wayne drank this water for about thirty years until more droughts made it less dependable. In 1931, the mayor, William Hosey, decided to build the Three Rivers Water Filtration Plant, which cost $2.5 million in the throes of the Great Depression. After a few expansions, the plant now has an underground reservoir that holds twenty million gallons of water. It is capable of pumping more than one hundred million gallons of water a day through six hundred miles of pipes. It pumps out eleven billion gallons each year before that water can even get to the Maumee.

The St. Joseph River Dam is not the only dam in the Maumee River system—not by a long shot. In fact, you only have to walk about three quarters of a mile to find the first one on the Maumee itself. But for me, the St. Joseph River Dam represents something fundamental about the relationship between people and the Maumee River: our uneasy entanglement. It is nearly impossible to talk about the history of the Maumee without addressing what has been done to it by humans; and I have come to believe that the reverse is also true—that any discussion of the history (and future) of life in this region is incomplete if it makes no mention of the Maumee River.

If you want to drive the length of the Maumee River—from its headwaters in Fort Wayne to its mouth near Toledo—you can do it in about two hours. It's a straight shot of just over a hundred miles. According to an estimate by Google, it would only take you thirty-three hours to walk that same distance. That is, if you are the kind of person who could walk for a day and a half straight through without eating or sleeping or passing out from exhaustion. I am not that kind of person. I am not even your standard wilderness explorer type. I spend a lot of time in fields and in the woods and in gardens, but I don't go on long backpacking trips or camp for weeks in the mountains. I don't fish. But when I was a kid, I would go to the forest near our house and sit down for a few hours to watch and sketch the trees and birds and rocks. I filled more than a few notebooks this way. On my first date with my now-wife, we sat in a big field and charted constellations. I compulsively acquire maps. At museums, I read every exhibit tag and every piece of available literature. What I am trying to say is that the trip would take me more than a day and a half straight through.

I decided to walk most of the way, but canoe a stretch on the lower half. I would walk alone, and my friend, Jason Bleijerveld, who is a frequent canoer and who also happens to own a canoe, would join me for two days on the water. The idea was to move slowly, nearer the river's pace. I decided to make the trip in early August, when the spring rains have long subsided and the stagnant heat of late summer begins to coax the algae into bloom. In the weeks leading up to my journey, I borrowed a bivy sack and tarp, called a couple of people about places to stay, bought a pack, three new maps, a fresh notebook, and a decent pair of shoes, and got a shorter haircut. And because I did not want to carry any of my large field guides, and because I like to know what things are called, I also picked up two slim, lightweight identification books:

Peterson's *First Guide to Birds of North America*, and *First Guide to Trees of North America*. I did not buy the guide to wildflowers, a decision which I regret.

A few days before I left, a local news headline declared: "Another Algal Bloom Found in Maumee River." The accompanying article went on to say that the bloom was in Defiance County, just about halfway from one end of the river to the other. The Defiance County Health Department mentioned that there had been a "green film" on the surface of the water, and that it was not clear what had caused the algae. The health department said that it did not know if the bloom was toxic or harmful, but that residents should avoid any water "that looks like spilled paint; has surface scum, mats, or films; is discolored or has colored streaks; or has green globs floating just below the surface." Nevertheless, the report assured readers that the drinking water in the city of Defiance was safe. It was only the river that was suspect.

On the eve of my departure, I felt a heightened alertness, an irrepressible mix of excitement and apprehension. I went to bed with my mind full and active. Sleep came fitfully. The next morning I woke early, fried a couple of eggs, pulled on my shoes, loaded my pack, and took off downstream.

CHAPTER 2

I t is not easy to see the headwaters of the Maumee. Probably the best view is from the Columbia Street bridge, which straddles the river just below the confluence. This is where I began my walk. To get there, I walked out along a narrow sidewalk toward the middle of the bridge, past vehicles that swept by in stoplight-regulated waves. I was wearing what would be my uniform for most of the journey: brown cutoff shorts rolled up just above the knees, a light grey pocket T-shirt, and wool socks pushed down around my ankles. On my head was a plain ball cap, navy with a green brim. From the midway point of the bridge, I could look northwest onto the headwaters. The water was wide, perhaps two hundred feet across. Ahead and to the left was the St. Marys, to the right the St. Joseph, and directly below me, the first lengths of the Maumee. This was the confluence, that no-man's-land where the rivers come together. But which parts, I wondered, belonged to which?

I pulled out my map and leaned in close to the section where the confluence formed. This was a detailed map, but it was still not entirely clear where one might draw the lines. The US Geological Survey is little help. Its guidelines for determining the length of a river are as follows: "The length may be considered to be the distance from the mouth to the most distant headwater source (irrespective of stream name) or from the mouth to the headwaters of the stream commonly identified as the source stream." By that measure, one could conceivably trace the origins of the Maumee all the way to the source of the St. Joseph, or the St. Marys. (Then, of course, the question becomes: where does the St. Marys River begin?) From above, as from the side, there are no clear divisions. The word confluence comes from the Latin *confluent*, meaning "flowing together." Together, and into each other, the rivers flow loosely across any lines that we might draw to separate them.

I turned back down the sidewalk and crossed the road, joining the Rivergreenway trail on the opposite side as the sun burned off the last remnants of morning haze. This was a familiar route, along the top of the embankment that contains the north side of the Maumee. By eight o'clock it was already hot. The air hung thick with water as I walked through it heading east. I kept adjusting my pack to accommodate the distribution of its contents. To my left, the grass spilled down and met a one-way street, across which was the neighborhood in which I used to live. To my right, the downslope met a rock berm of the kind that line human-made ponds and lakes—small boulders marking the boundaries of the river. Ahead, the Maumee poured out of sight, plunging around a bend and through the heart of the watershed. I followed.

———————————

When the last glacier withdrew from this country, it smothered the base layer of Paleozoic bedrock in a succession of moraines—ridges of rock and sediment laid along the edges of the receding ice. The most pronounced were near the modern-day cities of Fort Wayne and Defiance. Each is centered on the Maumee River, with its ends bending east like the lines in an italicized *v*. Stretched between them are the dirt and rocks that gathered underneath the glacier—known collectively as 'till'—laid out in either long, flat plains, or waves of gentle hills. In the millennia since, people have done their best to make an impression, but, geologically speaking, the Maumee Basin was written in ice.

The ice arrived twenty-two thousand years before the present. The most recent ice, that is—the Wisconsonian iteration of the Laurentide ice sheet. At its largest, the Laurentide sheet covered six million square miles, and contained 35 percent of the world's ice. It was centered over Hudson Bay in modern-day New York State, and stretched as far west as North Dakota. Each southward advance of the sheet involved a different projection, or "lobe," of ice. In the Maumee River Basin, these were, in order, the Huron-Erie Lobe, the Saginaw Lobe, and the Erie Lobe. The ice sheet came about, like most glaciers, during a period of excessive cold. Freezing snow and rain accumulated year over year, without a summer to melt. The resultant glacier spent ten thousand years molding ground, barely a blink in geological time. In this long view of prehistory, the ice advanced and receded like waves washing over a beach.

The last spur of ice to cover the Maumee River Basin— fourteen thousand years before anybody would call it the Maumee River Basin—was the Erie Lobe. It reached about as far south as Fort Wayne, and there was a lot of it. Some geologists think the ice was almost a mile thick. As it moved across the earth, the ice arranged and rearranged the landscape, carved valleys and ditches, riverbeds and lakes. Glaciers possess a kind of power that is incomprehensible in scale. As with other unfathomable phenomena, one early way of attempting to reckon their force was to invoke the divine: Louis Agassiz, the father of glacial geology, called them "God's great plough."

Most geologists seem to agree that what happened next went essentially like this: as the earth warmed, the Erie Lobe melted. Its southern boundary retreated northeast, toward what is now Lake Erie. Meltwater filled the area between the glacier and the Fort Wayne Moraine, forming a huge glacial lake called Lake Maumee. Lake Maumee drained to the southwest, through a small channel, into what is now the Wabash River. Pressed against the Fort Wayne moraine, the lake exerted greater and greater pressure. Eventually, the strain became too much and the lake "overtopped a sag," washing down in a flood of water known as the Maumee Torrent. This massive surge of water carved out the ground twenty-five miles south, through a mile-wide slice called the Wabash-Erie Channel.

I walked down that channel, turned left at a bend in the Maumee River, and crossed a two-lane road. A sign near the road said, "NOTICE: This water way receives stormwater runoff and combined sewage." Then it said, "Contact with the water could make you sick." I continued on the Rivergreenway path under cover of shade, the river to my right. The trail was primarily surrounded by trees and brush. To my left were, respectively, a neighborhood; two bridges; three large cement wastewater holding ponds; a golf course; the Fort Wayne Biosolids Facility, which makes lime and compost; corn fields; several runners and bicyclists; and a woman on roller blades. My elevation changed slowly from a few dozen yards above the water to just a couple of feet. On my right, the river was visible through a scrim of branches and leaves. More than once, I stopped and climbed down to watch the light glint on the surface of the water, to hear the river rushing past itself over a layer of smoothed-out stones.

It could have been a glacier that dug the initial course of the Maumee River. More likely it was a fledgling stream, slicing downhill in search of the retreating lake. Either way, the Maumee came from water—washing over earth, rippling year by year across the landscape, snaking its way downstream.

The Rivergreenway ended in New Haven, just past the Fort Wayne city limits. It was lunchtime, so I stopped at a diner called Rich's Café. A sign on the door announced that because the air conditioner had gone out two days before, and because the afternoon promised to be muggy and hot, the diner would be closing early. But it was still open when I got there. I ate a turkey sandwich and cup of melon and grapes. After lunch, I had a choice: I could stay on the south side of the river and walk by another golf course, or cut back up north. On a whim, I went north.

Rich's Café was right; it was muggy and hot. There was no wind. The only sound came from cars and from my shoes, crunching the gravel berm. Sometimes there was not much shoulder at all—especially around curves, I had just a thin strip of rubble between the road and metal guardrail. Vehicles hummed along at fifty-five miles per hour. My whole body would tense up as they gusted past. When there were no cars, I had nothing to do but walk and think. I wondered what the temperature was; how many steps I had taken; the life circumstances of a scowling driver; whether I had put enough sunscreen on my calves. I tried to count the rows of corn in a field, but became dizzy and stopped. Once or twice, I caught myself humming the chorus of a song I could not name.

Best anyone can tell, the first people arrived in the lower Great Lakes region more than thirteen thousand years ago, chasing the retreating Wisconsinan glacier. They were probably later to the Maumee River Basin, given that it was probably still underwater. These Paleo-Indians formed bands of hunters to follow roaming caribou. As the climate warmed, oak-dominated forests and deer arrived, replacing Pleistocene plants and mammals. Around the time the Maumee was carving its route across the empty lake bed, the Paleo-Indians had begun to shake out into more well-defined cliques. Anthropologists call these "band groups." At some point the Miami emerged as a distinct tribe. In the late 1660s, there was a Miami village at the mouth of the St. Joseph River—a different St. Joseph River than the one that feeds into the Maumee, located near present-day South Bend, Indiana. Later, a Miami village stood where Chicago is now. Around the early eighteenth century, the Miami migrated south and east into the watershed of the Wabash River and along the banks of the Maumee.

The Miami origin story begins: *mihtami myaamiaki nipinkonci saakaciweeciki*—"at first the Miamis came out of the water." The water is the St. Joseph River (*Saakiiweesiipiwi*) that empties into Lake Michigan. The place of emergence is called *Saakiiweeyonki*—"The Confluence." Moving south, they built a village where the *Kohciisa Siipiwi* and the *Maameewa Siipiiwa* poured into the *Taawaawa Siipiiwi*: it was called Kiihkayonki. The land in the Wabash and Maumee valleys in which they made their home is called *Myaamionki*, or "the place of the *Myaamia*." The Miami mark their tenure in these ancestral homelands "from time immemorial until the arrival of Europeans in North America."

———————

Late afternoon, I came to a pull-off on the north side of the road. A trailhead split the tree line a short way in. It looked like it might go toward the river, so I took it. The trail was rough and steep, and led down to an outcropping of stones that looked like a jagged half-moon. The river's color was the same as before, but it was much shallower here, maybe a foot or two deep. I could see rocks—glacial till?—on the bottom through a gloom of murk. The water eddied around sticks and rocks and leaves, wrinkling like the lines on an aged face. White bubbles like clumps of spittle floated across the surface. There was brush around the beach, and trash—napkins, bits of plastic food wrappers, a big tire stuck half out of the water. Ants, disturbed, bustled around my feet. A kingfisher clicked in a swooping motion across the water to my left.

I had crossed some time ago to the south side of the river, and this beach—if that is the right term for it—was along the bottom edge of a horseshoe curve, which bent north in either direction. The banks on either side rose sharply, and both behind me and on the opposite rim a few rows of trees lined the top, marking the alleged boundaries of an adjacent field. I leaned my pack against a rock and wiped my face with the front hem of my shirt. It was cooler by the water than up on the road, and I had not realized how sweaty I was. My shirt was soaked through around the shoulders in the outlines of pack straps, and the back, originally a medium grey, had turned completely dark. I shook it out to let my skin breathe, then sat back against a big tree root to rest.

Nobody knows who was the first white person to navigate the Maumee River. The first to the territory were probably French-Canadian fur traders. *Voyageurs.* These were rugged, wilderness characters who traveled around hunting and trapping. They are perhaps most famous for their birchbark canoes. These required constant maintenance, so each canoe carried extra bark, spruce root for sewing, and bags of pine resin for sealant. At night, voyageurs would tip a canoe over on its gunwale and stretch a canvas sheet from it to the ground, forming a tent. Voyageurs wore linen shirts with a long slit at the neck, and sashes around their waists. On their heads were a mix of fur caps, brimmed felt hats, and long, brimless knit hats called toques.

Other early arrivals were Jesuit priests, members of a Catholic order intent on bringing the Christian gospel to new people groups, which included Native tribes. They created the first dictionaries of many Native languages for that purpose. Some of these still survive. People working to reconstruct the Miami-Illinois language—which was functionally dead for much of the twentieth century—have been using a 580-page French Jesuit manuscript from the late 1600s known as the Gravier Dictionary. Jesuits traveled all over the place spreading their religious message, and were responsible for some of the earliest maps of the region, charting landmarks, Native settlements, military outposts, and rivers.

It was probably a priest who gave the rivers their European names. There is a statue of this anonymous priest at the headwaters of the Maumee, near the St. Joseph River Dam. It is a bronze statue, turned green by oxidation, showing a stern-looking man in a long tunic. He has a sharp nose, a close bronze beard, and bronze hair that hangs down over his ears. He is facing the confluence with one arm stretched out as if he is about to say something dramatic. Maybe the statue is supposed to look like Father Joseph Pierre de Bonnecamps, a Jesuit Priest, cartographer, mathematician, and professor of hydrography at the Jesuit University of Quebec, who traveled with the explorer Pierre-Joseph Celeron de Blainville in 1749. He is the most likely candidate for the priest who named the rivers. De Bonnecamps drew a map that included Lake Erie, the Ohio River, the St. Marys River, and the Maumee—though, strangely, he left the latter unlabeled. Of the area immediate to Lake Erie on the south and west, including much of the Maumee Valley, de Bonnecamps wrote, *"Toutte celle partie du lac est inconnue"*— "All that part of the lake is unknown."

I wasn't the only one to confuse the boundaries of the Maumee. Early mapmakers did too. Some of them thought it was the Ohio River. Others thought the Maumee was part of the Wabash. The earliest maps of the region were drawn by Frenchmen in the seventeenth century, who seemed to miss the headwaters altogether, maybe because they were sticking mostly to the lakes. The explorer Louis Joliet drew a map in 1672 on which the Maumee was distinct from the Ohio, but which showed the two running extremely close together, when they are in reality no less than one hundred and fifty miles apart. One map from 1720 shows a single long river, labeled *Ouabache Autrempt Appellee Ohio ou Belle Riviere*, running from Lake Erie all the way to the Mississippi. Another map, from 1742, shows the Wabash and the Maumee running roughly parallel to each other. By 1794, the settlers had pretty well figured it out.

Europeans began to arrive to the area in significant numbers during the seventeenth century. Maps were not only useful for navigation; they were also a chance to draw boundaries and claim land. This was the reason Celeron de Blainville and Father de Bonnecamps were on the Ohio River in the first place—to lay claim to its valley for the French. De Blainville buried lead plates along the Ohio and its tributaries to mark the new territorial boundary. He put one of these plates at the junction of the Kanahwa and Ohio Rivers, in what is now West Virginia. It read:

> In the year of 1749, of the reign of Louis the 15th, King of France, we Céloron, commander of a detachment sent by Monsieur the Marquis de la Gallissonieré, Governor General of New France, to reestablish tranquility in some Indian villages in these provinces, have buried this plate at the mouth of the River Chinodahichiltha on the 18th of August near the River Ohio, otherwise Beautiful River, as a monument of the renewal of the possession we have taken of the said River Ohio, and of all those which empty into it, and of all the lands on both sides as far as the sources of said rivers, as enjoyed or ought to have been enjoyed by the kings of France preceding, and as they have there maintained themselves by arms and by treaties, especially those of Ryswick, Utrecht, and Aix la Chapelle.

After laying this marker, de Blainville and company traveled west down the Ohio River, along what is now the southern border of Ohio. They buried the last of the lead plates at *Riviere a la Roche,* near modern-day Cincinnati, then turned north, toward the headwaters of the Maumee.

Maybe the most violent boundaries in the region were drawn at the Treaty of Greenville in 1795. This was a treaty between the United States and the Native tribes who had previously controlled the region, including the Miami. The Miami fought on the side of the British during the American Revolution, under the command of Mihsihkinaahkwa, or Little Turtle. Not much is known about Mihsihkinaahkwa as a person. In *The Life and Times of Little Turtle,* historian Harvey Lewis Carter points out that the available sources of information on him are almost entirely through the eyes of others—namely the US government, his enemies. The artist Gilbert Stuart did paint a portrait of Mihsihkinaahkwa in later years, but it was destroyed when the White House was torched in 1814; a surviving lithograph shows a man with a prominent nose, high cheekbones, soft eyes, and a large forehead, smiling slightly.

It is generally believed that Mihsihkinaahkwa was born sometime around 1752 on the Eel River, that his father was a Miami chief and his mother a Mahican, and that he was tall and had a quick wit. (Once, when asked by the French writer Volney whether he thought that his people might have been descended from the Tartars, given an apparent similarity in facial features, Mihsihkinaahkwa responded, through an interpreter, "Why not think the Tartars descended from the Indians if they look so much alike?") He was also regarded as a brilliant strategist, which made him a natural choice for war chief. It was Mihsihkinaahkwa who became friendly with the British and decided to help them out fighting the United States.

After the Revolutionary War, it turned out that the Miami, as well as some other tribes like the Shawnee, were right in the middle of the Northwest Territory, which had been signed over to the US by the British without so much as consulting the people who lived there. The Miami were now "in the way," so to speak. After what historian Michael Warner calls "several years of smoldering hostilities and mutual provocations," the newly-formed United States went on the offensive, planning a series of attacks on Native towns along the river in the Maumee Valley. General Joseph Harmar of the US Army would take the largest force and march on Kiihkayonki.

Harmar's military career had been anything but spectacular thus far, and was about to get worse. In October of 1790, his army arrived at the village only to find that Little Turtle and the rest of the residents, including British and French traders, had left. They spent the next few days near the confluence of the St. Marys, the St. Joseph, and the Maumee, burning corn and vegetable stores and looting homes. A small group of soldiers, led by Colonel John Hardin, wandered up north and were ambushed. They still managed somehow to carry on the business of torching wiccias, and, having finished this task, left the village on October 21 to head back to Fort Washington, near what is now Cincinnati, Ohio. Hardin was not happy about being ambushed. He tried to convince Harmar that they should double back and attack the residents of former Kekionga as they returned to the coals of their village. After some back and forth, Harmar agreed.

On October 22, the leaders each took a contingent of soldiers and began to surround the Miami capitol. Harmar wrote in his journal that the following day was warm and sunny. Autumn is a beautiful time in this country, and especially late October. The air is crisp, and the sun, which in summer burns hot and dry, takes on a lighter, golden warmth. By October, the leaves have begun to turn. Harmar was no doubt writing among trees which had only recently changed suddenly and theatrically, as if lit by a match, to hues of yellow, orange, and red.

Then, a shot rang out. Maybe a flurry of shots. The Miami took off running, and soldiers charged after them across a ford in the Maumee. It's worth mentioning here that Harmar's band of soldiers could only generously be called an army. More accurately, it was a loose bunch of undisciplined men with little to no military experience. Some of them had never even shot a musket. In the din of battle, they acted rashly. The Miami, led by Mihsihkinaahkwa, took their stand a mile or so north, near a French fort, and killed more than two hundred members of a fragmented US. Army. Hardin's men fled the scene, leaving the bodies of their dead lying in the shallows of the river alongside those of the slain Miami.

The Battle of Kekionga, or the Battle of the Maumee, as it is sometimes called, was the beginning of the Northwest Indian War. A year later, President George Washington sent General Arthur St. Clair—who was also the governor of the Northwest Territory— back to the area to clear out Native tribes and smooth the way for European settlement. St. Clair went, but his army was routed by a confederation of warriors led by Mihsihkinaahkwa, Weyapiersenwah (or Blue Jacket) of the Shawnee, and Buckongahelas of the Delaware. Mihsihkinaahkwa was the primary strategist. The confederation attacked St. Clair's forces in the blue light of dawn and killed more than six hundred soldiers. This battle, fought near the banks of the nearby Wabash River, was the greatest defeat dealt to the US by any Native nation, with more than three times as many casualties as Custer's famous last stand.

A few years later, General Anthony Wayne—who earned the name "Mad" Anthony Wayne for his hot temper and fiery disposition, and probably also his wild and impulsive battlefield behavior—mounted a crusade against Native tribes in the region. During this time, Little Turtle's white son-in-law, William Wells, who had been kidnapped by the Miami as a child and formally adopted into the tribe, and who had grown up to marry Mihsihkinaahkwa's daughter, either joined the US Army, or was captured by them, depending on who you ask. Wells became General Wayne's chief of scouts, and Wayne's rampage continued, ending in a decisive victory on the other end of the Maumee River near what would become Toledo, Ohio: the Battle of Fallen Timbers.

Mihsihkinaahkwa decided then that resistance would only cause greater harm to his people. At one gathering of chiefs, which included Blue Jacket and Buckongahelas, Little Turtle spoke out. Wearing "foot-long ear rings," and large jewels in his nose, he stood in the light of a fire and spoke. Of the United States, he said: "We have beaten the enemy twice, under separate commanders. We cannot expect the same good fortune to always attend us. The Americans are now led by a chief who never sleeps; the night and the day are alike to him. And during all the time that he has been marching upon our villages, notwithstanding the watchfulness of our young men, we have never been able to surprise him. Think well of it. There is something whispers me, it would be prudent to listen to his offers of peace." (Of course, he was not speaking English.)

On August 3, 1795, Little Turtle signed the Treaty of Greenville beside ninety-two other tribal leaders. General Anthony Wayne signed first and then each chief drew a symbol which represented his name, beside which a notary wrote its phonetic spelling using English letters. There were 1,130 Native people at the signing in total, seventy-three of them Miami. There were eight sworn interpreters, including Little Turtle's son-in-law, William Wells. The treaty promised "to put an end to a destructive war, to settle all controversies, and to restore harmony and friendly intercourse between the said United States and Indian tribes." This was to happen on General Wayne's terms, which meant handing over some of the Miami lands—including the headwaters of the Maumee River. Mihsihkinaahkwa bristled at the demands the United States was making of the tribes. He reminded Wayne that this land had always belonged to the Miami, and that he hoped they would be able to persist there. Then he signed his name.

Maybe Little Turtle overestimated the goodwill of his conquerors, or maybe he didn't have much of a choice, but the steady encroachment of U.S. citizens onto areas reserved as Indian Territory would continue for the next four decades. And not quite fifty years later, after the US Congress signed the Indian Removal Act, hundreds of Miami were forcefully removed from their homelands by the United States. They were loaded onto canal boats and shipped through Fort Wayne to Defiance, south to Cincinnati, then west on the Ohio River—first to Kansas, and then again, years later, to Oklahoma. But the Miami were not altogether gone from this place. Some stayed behind during the removal, and others moved back. Those who stayed in Indiana formed the Miami Nation of Indiana. Those who ended up in Oklahoma organized themselves as the Miami Tribe of Oklahoma. The Miami Tribe of Oklahoma is the only federally-recognized Miami group. The Miami Nation of Indiana has yet to be recognized by the United States government.

In 2005, several members of the Miami Tribe of Oklahoma co-authored a paper on the removal of the Miami from their homelands. It began like this:

> For many Myaamia people, we often say that time is like a pond, and events are like stones dropped in water. Emotionally powerful events create big ripples that combine with the smaller ripples of less powerful events in unpredictable ways. In our Myaamia pond, the forced removal was more like a boulder, which, once dropped into our lives, created a series of waves that changed everything. The political, economic, and cultural impacts of this forced relocation were immense, and the emotional toll of this experience has trickled downstream in the memories and stories of many Myaamia families. Removal remains an event that is painful for us to remember and discuss, but to choose to forget has never been an option. We know we must continue to remember in order to honor the sacrifices endured by our ancestors who made this terrible journey.

Myaamia, or sometimes Myaamiaki, is what members of the tribe call themselves. Translated into English, it means "the downstream people."

———————

Growing up in Fort Wayne, I knew about Little Turtle. One branch of the local library is named after him. When I was ten years old, my mother drove me, with my brother and sisters, to the war chief's grave near the junction of the three rivers. It's tucked between backyards on a quiet residential street a quarter mile or less from the confluence. To mark the grave there is a rectangular bronze plaque embedded in a flat grey stone. The plaque is worn on the edges, buffed by the years, and looks almost polished in places. It reads:

CHIEF LITTLE TURTLE
1752-1812

ME-SHE-KIN-NO-QUAH, CHIEF
OF THE MIAMI INDIANS
TEACHER OF HIS PEOPLE
FRIEND OF THE UNITED STATES

HIS ENDEAVORS TOWARD PEACE
SHOULD SERVE AS AN INSPIRATION
FOR FUTURE GENERATIONS
THIS PLOT OF GROUND, THE LAST
RESTING PLACE OF CHIEF LITTLE
TURTLE, IS DEDICATED TO THE
CHILDREN OF AMERICA AND MADE
A PUBLIC PARK IN 1959
THROUGH THE GENEROSITY OF

ELEANOR SMELTZLY AND
MARY CATHERINE SMELTZLY

UNDER THE AUSPICES OF THE
ALLEN COUNTY–FORT WAYNE
HISTORICAL SOCIETY

Mihsihkinaahkwa is a complicated character. So far as I can tell, he loved his people and the place in which they lived. He always opposed the intrusion of white customs and values into Miami life, and felt that the United States was too land hungry. But he was so concerned about the threat of US violence that he signed three more treaties after Greenville ceding land to the new country. Some historians think the tribe lost respect for Little Turtle after he signed the Treaty of Greenville, though he continued to live among them until he died in 1812. At the same time, Little Turtle's compliance made him quite popular with US leaders—he toured the country later in his life, and was even taken to Washington to meet the president, who gave him an ornamental sword.

Returning to Little Turtle's plaque years later, when I lived along the Maumee, I began to understand the trouble with its claims. There is a certain amount of arrogance required to declare Mihsihkinaahkwa a "Friend of the United States," whose "endeavors toward peace should serve as an inspiration for future generations," after destroying his community and using military power to pressure him into compliance. Or maybe it is stubborn ignorance, that most necessary quality for a society hell-bent on imposing itself at any cost. Whatever the reason, for residents of Fort Wayne, Mihsihkinaahkwa, like the rivers, is obscured; it is difficult for us to see him.

I pulled on my pack and returned to the road. Clouds traced across the sky, dulling the sun's light but not its heat. Around the first bend the river cut away left, taking the trees with it, and I was soon flanked by fields of grass lying lower than the pavement on either side. I moved from the thin gravel berm down into the grass on the left-hand side of the road, glad to put some space between myself and the cars that did not move over when they passed. On the opposite side of the road, and about a hundred yards back, was a tire factory. The factory was almost half a mile long, low and white, with blue accents and, at one end, a large American flag hanging on a pole in front of a row of tinted glass windows. A semi pulled out of the drive and onto the road nearly every minute, like clockwork.

Around five o'clock I came to a cut-in with a view through thick trees. I stumbled down a thin trail to the river below—attempting, with limited success, to use exposed roots like steps on a flight of stairs—and found the water's edge, with a small indent in the bank where a layer of green scum had accumulated. I took some photos to remember it, then climbed back up the trail. Just past the overlook was a small cemetery, perhaps two hundred graves. These were mostly simple in design, and arranged in neat rows that ran perpendicular between the river and the road. I could read each stone clearly just steps away: Becker, Gerig, Wheeler, Hall. A stone marker at one end read: EV. MENNONITE CEMETERY. The cemetery was established in 1901, 110 years too late for the Miami that died in the river upstream, but within the lifetime of the Maumee, which had watched through the trees as each of their bodies was returned to earth—as one after another they were laid to rest. I have always thought there was a certain solemnity to rivers, and it occurred to me that this might have to do with the number of deaths they have seen.

Walking past the names in the cemetery, I thought of the Miami, the residents of Kiihkayonki; I thought of the early voyageurs, the Jesuits, and Jean-Pierre de Bonnecamps; I thought of Mihsihkinaahkwa, and his other name, Little Turtle; I thought of the names of the dead who had lain in this river some one hundred years before, though I did not know them. There was a lull in traffic on the road, and the air was quiet except for a breeze nudging at the grass between graves. I felt that I should do something, so I took off my hat and said a simple liturgy that I remembered from time spent with a Mennonite church: "God, have mercy." I stood still for a moment and then continued east. Once I had passed the edge of the cemetery, I put my hat back on my head.

CHAPTER 3

Heading eastward from Fort Wayne, the Maumee River winds its way in long, sinuous curves across the Ohio border and into the northwest corner of that state. The official term for these curves is meander loops. Some people also call them switchbacks. An s-curve, or two meander loops back to back facing different directions, is called a meander. A whole string of meander loops laid out one after another along the course of a river is called a meander belt. On the Maumee, as with most rivers, meander loops look to be haphazardly arranged. Their basic process is a kind of turbulence that forms in the natural curves and indentations in a stream, known as helicoidal flow. The water moves like a sideways corkscrew: coming around a bend, it hits the river's outer bank and wears it away, then moves down along the bottom of the river, carrying chunks of the riverbank with it and dropping them on the other side. The inner bank builds inward as the outer bank is eroding outward, and the whole river starts to curve. This is called lateral migration. If the loop migrates out too far, and takes too much energy to travel across, the river will revolt and cut a shorter path, abandoning the meander loop, building up a sediment wall, and leaving standalone former riverbeds of water called oxbow lakes.

One consequence of the river's meandering course was that I could not follow it exactly. There is a lot of private property along the Maumee in this area, and most of it extends all the way down to the river. I had seen several signs that warned trespassers of a property owner who would shoot before asking questions, and though I had begun to have mixed feelings about the idea of a person owning a riverbank, I wasn't yet ready to be shot for them. So I stuck mostly to old US 24, which has a public right of way and which also follows the course of the river fairly closely. The river, for its part, came near the road often enough to assure me that it was still there. I could follow its course visually, even from the road, because it was flanked by a tall, dense tree line that wound with it on my left. As I walked, this tree line would swing in close and then open back out wide with each loop, coursing around a field or clearing before returning near the road.

I had permission to camp that first night in a forest preserve seventeen miles downriver of the confluence. About half a mile in I found a clear spot near a ledge that looked out over the river. I set up my tent, then wandered around the trails. The trees— yellow poplar, red oak, shagbark hickory, black walnut—were a deep green at this point in the season. Off-trail was a tangle of understory. Once or twice a minute, a walnut fell through and hit the ground with a plunk. The walnut, which is dark brown and ribbed, falls encased in a rough, lime green husk just larger than a golf ball. I picked up a few and chucked them as far as I could into the woods. Back at the campsite, dinner was peanut butter and honey wraps and a green apple almost the exact color of a walnut hull. I walked out along a ledge about fifteen feet and sat down on a thin strip of grass to look over the river. Below the ledge was a steep, brush-covered slope that abutted the water twenty or thirty yards down. I spent most of the evening there, with my back against a tree and my legs hanging over the edge, watching the river move past.

The next morning I walked three hours across the state line on Highway 424, toward Antwerp, Ohio. Most of the fields I passed on either side were corn. I went down and stood next to a row of stalks that beat my height by a foot. Some of the fields had soybeans that came to the middle of my shins. As I walked along one pasture, three cows stopped their grazing to amble over and get a closer look at me. Beyond the cows were about a hundred round bales of browning hay rolled up next to the road two deep. The prettiest fields were full of long grasses and edged with purple clover and Queen Anne's Lace. One field held a swarm of other flowers which I did not recognize, and which I could not look up because I had not brought the wildflower field guide.

Around the edges of some of the fields were swaths of forest. At one unkempt field without a "NO TRESPASSING" sign, I veered in. There was at least some chance this property belonged to no one in particular, I reasoned. The treeline sliced away from the road—roughly the river's course, though the trees were thick enough to mostly obscure my view. I did not mind. Trees are almost unanimously spectacular, and like snowflakes, singularly so. Halfway around the meander loop, I got out my field guide and began to stare at bark and leaves, trying to figure out what variety of trees these were. I wrote their names down in my notebook:

oak (easy; maybe red)
black walnut
silver maple (lots)
butternut
American crabapple?
pawpaw? (maybe)
some version of sumac
yellow poplar
hickory
beech

It used to be that the Maumee ran along the edge of an enormous wetland called the Black Swamp, or sometimes the Great Black Swamp. As recently as the early 1800s, marsh and thick forest covered at least fifteen hundred square miles—an area slightly larger than the state of Rhode Island. Some say it was two or three times that size. The swamp forest held white oak, beech, cottonwood, red oak, elm, black ash, white ash, blue ash, black walnut, hickory, sycamore, black willow, trembling aspen, basswood, shingle oak, chestnut, black cherry, mulberry, ironwood, sugar maple, soft maple, dogwood, buckeye, wild plum, thorn, pin oak, burr oak, prickly ash, honey locust, pawpaw, wahoo, water beech, slippery elm, hackberry, judas tree, apple, and June berry. Trees grew close together, tall and straight. The soil was largely clay, which kept water from absorbing into the ground, and the resultant muck was thick and deep.

The Black Swamp was known as one of the most fearsome stretches of land in the territory. Even the Native tribes living around the area opted to spend their time elsewhere, using the swamp primarily for hunting and fishing. The whole place was dark and dank. Mosquitos and insects were large and swarmed in thick clouds. The swamp was a tangle of mud and trunks—all but impassable on foot, and not much easier on a horse. According to one account it would take something like three days to go a distance of twenty miles, as "man and horse had to travel mid-leg deep in mud." And forget about wagons. A Moravian missionary named Michael Zeisberger wrote of "deep swamps and troublesome marshes…where no bit of dry land was to be seen." It was so difficult to find a dry spot for camp that some people, wading through on horseback, elected to spend the night sitting up in their saddles, leaning against a tree for stability.

Predictably, nobody wanted to live there. Those who did try to settle near the swamp were often sick from infections and diseases—erysipelas, an infection that caused painful, swollen lesions, chills, and vomiting; "ship fever," probably a form of the bacterial disease typhus; "the shivers," no doubt exactly what it sounds like; and a malarial deviant called "Maumee fever." A catholic bishop named Louis de Goesbriand visited Toledo in the early 1800s. He wrote later that "The Maumee Valley at this time was literally a land which devoured its inhabitants. The Maumee fever spared no one; the disease slowly but surely undermined the strongest constitutions, and there was not an old man to see in all that country." This was not an attractive prospect to homesteaders, and so most people either stopped in the southern part of Ohio or continued through to Michigan.

The first official road through the Great Black Swamp was approved by Congress in 1823, and completed in 1827. It soon gained a reputation as the worst road on the continent. The Maumee and Western Reserve Road was essentially a clearing 120 feet wide, low and wet. Trees on either side were one hundred feet tall, and blocked out most of the sun. The ground was somewhere between swamp and dirt. The more it was traveled, the worse it became. The road crossed twenty-two streams through the Black Swamp, and these would back up, flooding it with water. In the early 1830s, some people set up shop hauling stalled wagon teams out of the mud along the road. There was one alternative route: a trace cut several years earlier by an army under the command of a General Hull. Hull's Trace—it could hardly be called a road—crossed the Black Swamp further to the west, was barely the width of a set of wagon tracks, and somehow managed to pass through terrain worse than the Maumee and Western Reserve Road.

I walked the rest of the loop and back out onto Highway 424, a road settlers would have killed for—all sunshine and solid pavement. About a mile past the state line, a speckled, grey-black mutt without a collar showed up. She trotted contentedly alongside me for half a mile or so, tongue flapping out one side of her mouth, before she stopped and turned around to go back wherever she had come from.

———————

Antwerp, Ohio, is tucked inside a meander loop on the Maumee River three miles past the Indiana-Ohio border. It was platted in 1843, when the state of Ohio was already forty years old. After the town had been laid out, the surveyors looked in the post office directory to find a name that wasn't already being used, and "Antwerp" was one. A sign that I passed on the outskirts of town said: "Welcome to Antwerp: The Diamond of Northwest Ohio." In town, I passed a giant grain elevator and its associated grain bins, a Subway, a gas station, and a hair salon, then came to an intersection that had two handmade signs pointing opposite directions. One said "Garage Sale" and the other said "Huge Sale." I followed the sign for "Huge Sale" a couple of blocks down to a sale that was not huge. The blister on my heel was making its presence known. I had passed a pleasant looking coffee shop on the way, and since it was lunchtime, I went back there for something to eat.

The coffee shop was called The Agora on Waldo's Hill. It was in an old house with red wood siding and a large wraparound porch. The house had been retrofitted with a counter, tables, and inspirational craft décor. The barista, whose name was Aaron, did not know Waldo, but I found newspaper clippings on the wall near the bathroom that told me the hill was named for Waldo Witt, a local custodian and grain tester who enjoyed cigars and going to bed early. Waldo had died in 1998. When the hill was purchased years later, its new owners, who began the coffee shop, decided to keep his name alive. I returned to the counter, ordered a Turkey Reuben and a glass of water, then sat down inside at a table near the windows.

Aaron the barista was also Aaron the cook, and he began to assemble my sandwich. Aaron had medium blond hair cut short on the sides and combed over on top. His face was wholesome, his voice high and clear. He impressed me as the kind of person who tries to laugh as much as possible. Jolly is a word that came to mind. Aaron wore a red T-shirt that said "Live Generously" in white lettering, and sang along with the music playing from a speaker in the corner of the room. There were only two other customers and they were sitting outside on the porch, so I asked Aaron how long he had worked at The Agora on Waldo's Hill. I learned that he was a junior at Kent State who was back in town for the summer, that he grew up in Antwerp, that he had worked at the coffee shop off and on since before he left for college, and that he was not shy. When I told him I was walking through town his face screwed up into a question mark. "What would make you backpack through Paulding County?"

I said that I was traveling the length of the Maumee River and planned to write about it. Then I asked what he knew about the river.

"It's really shallow around here," he said. "Unless you get a lot of rain like last year, when it flooded all the farmers' fields. I used to canoe it with the Scouts, and it would get annoying how shallow it was. But there were lots of islands to stop and eat lunch on. It does flood sometimes though, and that's bad for the farmers." He considered for a moment. "Hmm, what else? They found a body in the river this spring, I think over near Indiana. We've had four murders in Antwerp so far this year, but they caught the guy."

Then he abruptly switched subjects. "Have you been to Doc Bricker's?"

I said I hadn't, and he gestured behind him. "Over here on the river is this huge house, Doc Bricker's. He was the dentist, so he could afford it. But he moved to Ethiopia or somewhere. He was a big game hunter, but he worked with local tribes so he would give them the meat and it wasn't a waste of an animal. He had them all over his house."

"The animals?"

"Yeah. Taxidermied. There was a huge hippo head in his basement. The schools used to take kids over there on trips, and there was an elephant head in the window that could really scare you. We used to canoe by it."

At this point a couple of customers came in and Aaron began to wait on them. I finished my Reuben. It was early afternoon and I had no clear idea of where to sleep for the evening. My plan had been to stay with a friend of a friend, but those arrangements had fallen through and I had not been able to find any hotels or campgrounds near Antwerp. The sign outside of The Agora advertised lodgings in addition to food and drinks, so during another lull in business I asked Aaron about this.

"It's an apartment upstairs that we rent out. But we're actually booked right now, I think up through next month. And there aren't a whole lot of other places around here. Nothing in walking distance, anyway." He paused. "Although—if you have your own tent and everything, you could probably camp in my parents' backyard. We live close, on the edge of town. Let me just give them a call and ask if it's okay."

Aaron called his mother and explained the situation. After a short conversation during which I did my best not to eavesdrop, he pulled the phone away from his ear, leaned toward me, and said, "She wants to know if you're a serial killer."

"No," I said. "I'm not a serial killer."

He repeated my answer into the phone, then turned to me again—attempting unsuccessfully to hide his amusement—"She says that's exactly what a serial killer would say."

After Aaron's shift ended at the coffee shop, we drove out to his parents' house in an old Ford Windstar that belonged to his sister. Aaron had brought four iced coffees with him that I held in my lap as we bounced through town. When we arrived, he introduced me to his girlfriend, his sister, and his sister's husband, and I handed out coffees. They were preparing for a weeklong trip to Florida, so I thanked Aaron for his help and excused myself to give them some space. Then I walked the three-quarters of a mile back into town to poke around.

Cutting north across the railroad tracks, I saw the police station, two churches, and a whole row of empty storefronts; read a sign in the window of City Hall that advertised the Ehrhart Museum of local natural history; went inside to see the museum and found it closed; went to a park and walked shoeless through the grass; meandered down a couple of side streets, past where I guessed Doc Bricker's would have been, though I did not find it; sat for a while on a bench outside a grocery called Home Town Pantry; and inspected the grain elevator, which straddled the railroad tracks and loomed enormous over the town. The grain elevator and its bins were the dominant features of downtown Antwerp. I never saw a train pass between them.

Standing underneath the grain elevator, I spotted an Ohio Historical Marker in front of a building on a side street nearby. I went over to look at it. The building was a VFW hall, and the marker put the story of Paulding County's settlement like this: "Early pioneers subdued the massive forests that once formed the 'Black Swamp' and built a thriving city." In this single, unassuming sentence is contained the story of one of the most dramatic encounters ever to play out between humans and nature in the Old Northwest Territory.

It is no surprise that a wetland the size of the Great Black Swamp was an unwelcome obstruction for those trying to colonize the region. Two of the biggest reasons the people of the United States wanted to get into the swamp—besides the fact that it took up thousands of square miles of land in the middle of the territory—were its rich soil and enormous supply of timber. There was simply too much opportunity there to leave it alone. The trees would have to go first, but they would need something better than a glorified mud path to get enough people into the swamp and carry out all that lumber. And the shallow, meandering Maumee was inhospitable to barges. So they built a canal.

The Wabash and Erie Canal was completed in 1843. There were actually two canals along the Maumee. The Miami and Erie Canal started at the mouth of the river, at Toledo, running west to Junction, just past the city of Defiance, then cut south toward Cincinnati. The Wabash and Erie Canal met up with the Miami and Erie at Junction, then ran west, up the Maumee, through Antwerp and across the state line to Fort Wayne, across the Fort Wayne portage to the Wabash, then down the Wabash all the way to Lafayette, Indiana. (Eventually its westernmost point went as far as Evansville, on the Ohio River.) Antwerp was a canal town, built for its proximity to water as much as anything.

Workers broke ground on the Miami and Erie Canal in 1834. The Wabash and Erie began in Indiana in 1832, though it wasn't until 1837 that the contract was won for the stretch between Junction and the Indiana/Ohio line. Once that happened, two thousand laborers were hired to start building. Much of the way was still swamp. Conditions were filthy. Workers lived in crowded tents. Epidemics of smallpox, typhoid, and pneumonia broke out, and whole gangs of men would go down sick at once. This delayed construction of the canals considerably. There were other slowdowns, including a boundary dispute with the state of Michigan. In 1840, there were three full months during which nothing got done. Slowly, over the next half-dozen years, huge locks were constructed of stone and wood. Both canals ran alongside the Maumee, and water from the river was pumped in to fill them. The canals opened for traffic from Toledo to Fort Wayne on May 8, 1843.

On July 4, 1843, Fort Wayne held a jamboree. It had everything: a marching band, a mile-long parade, celebratory cannon fire, and a reading of the Declaration of Independence. That was probably the biggest Fourth of July celebration ever to take place in the city. Independence! They had carved their own river!

It was a momentous achievement, if short lived. The canals had taken so long to construct that they were outdated almost immediately. In 1854, the Ohio and Indiana Railroad reached Fort Wayne; in 1855, the Toledo, Wabash, and Western Railroad cut through Paulding County. Railroads could transport people and goods faster than the canals, and did not have to stick close to a water source like the Maumee. By 1857, the canals were already obsolete. In 1863 and 1864, two sections of the Wabash and Erie Canal from Junction to Fort Wayne were formally abandoned and drained. Antwerp, though it would eventually receive a railroad and a grain elevator, was orphaned; a canal town without a canal.

The canals marked the beginning of the end for the legendary forests of the Great Black Swamp. Towns and villages cropped up along the canal to house workmen who set about cutting and shipping timber. Some of the trees became homes and fires for the settlers themselves; others were sent east for shipbuilding. Some were burned into charcoal at Antwerp for smelting factories in Toledo. This went on for decades. By 1870, half the trees in the basin would be gone.

But there was still the matter of the soil. A man named William Woodbridge had noticed in 1815—nineteen years before the digging of the canals—that the swamp consisted of "a basin of hard clay" covered in "thick stratum of the most fertile black loam." Agriculture was in that loam's future, if political leaders had anything to say about it. An 1822 report to Congress observed that the swamp could be "made subservient to all the purposes of profitable agriculture." Lewis Cass, governor of the Michigan Territory, described "rich bottoms denuded of timber, as though inviting the labor and enterprise of the settler." (Translation: the land was asking for it.) Still, between the settlers and that soil were several feet of water. Farmers started plowing their fields into rows with furrows between, and digging long ditches between those fields to drain off water—often into the streams that fed the Maumee River.

In 1835, four hundred miles away in Geneva, New York, a farmer named John Johnson dug a series of these drains across his 300-acre farm. But he did something else which was not so common—lining his drains with clay tiles to create a channel that would keep the water from seeping back into the ground. It worked. The practice spread slowly. The first of these tiles was not laid in Indiana until 1850, and it wasn't until the 1870s that things really got underway. Then they exploded. By 1882, there was enough demand to warrant a thousand tile factories in the wetlands of Indiana, Illinois, and Ohio alone. Somebody realized that there were massive deposits of clay underneath this midwestern land, and production increased. Giant mechanical diggers with scoop-covered wheels, recently developed, dug ditches faster than fifteen men; engineers designed a series of canals to divert water from the surface, and machines gouged long ditches through the landscape. The soil of the Great Black Swamp was dried out using a complex series of underground tunnels, pipes, and clay tiling.

It hardly seems necessary to make the explicit connection between this activity and the widespread cultural belief in Manifest Destiny, which motivated, or at least justified, westward expansion and the removal of Native populations in the United States, but I am going to do it anyway. This justification, such as it was, had to do with the idea that the Americans, by virtue of their self-described moral and cultural superiority (and, not incidentally, their demonstrated military superiority), had been given a divine right to settle the continent and to do with it, and its previous inhabitants, as they pleased. Implicit in Manifest Destiny is the assumption that the capacity to do something is justification enough to do it. So the Native residents were killed or shipped west, and the forests were cleared, and canals were built, and the swamp was drained, all in the interest of progress and with astonishing indifference to the consequences. And thanks to the Industrial Revolution, settlers could perform this indifference with greater power.

I poked around downtown for a couple of hours, then went and ate dinner at Antwerp's only bar, the Oasis, a storefront bar with neon beer signs in the windows. A long counter ran the length of the room. This counter held, from front to back, the cash register, the bar, the kitchen, and a snack station. Above the bar was a brown menu board with white clip-in letters. Across from the bar was a row of booths with red faux-leather seats, and there were tables down the middle of the room holding a scattering of diners and drinkers. I sat down in a booth and ordered the signature sandwich—a burger—and a glass of water. My pack, sitting next to me, attracted attention. A couple of people walked by staring. The waitress, a friendly woman with a large smile, asked politely about where I was going, and when I said I was backpacking down the Maumee River she seemed pleased. One person, an older, frail-looking woman with silver hair and a wood-handled cane, stopped on her way out the door. She asked my destination, and how long I had been on the road. I told her that I was headed to Toledo and that it would take a little more than a week, all told. "Well," she said, "You'll probably get pretty lonely."

After dinner, I went back to Aaron's house and met his parents, Kimberly and Michael Schneider. Michael was a tall man with wide eyes, a large forehead, and unruly grey hair that wisped out to the back. Kimberly had short-cropped blond hair and thin-rimmed glasses, and wore a blouse that was speckled white, blue, and light blue. They offered the same uncalculating friendliness Aaron had. I thanked them for their hospitality, and Kimberly said, "Of course! My husband is a Methodist minister and this is a parsonage, so we like to think of it as God's property." Besides, she added, "You're not the worst person Aaron's ever brought into the house. Once he let a drunk in the back door in the middle of the night. The man wasn't even wearing shoes!" She laughed, high and giggling.

I pitched my bivy sack in the middle of the backyard, away from any trees, before going inside to wash up. Michael offered me a glass of water from a dispenser in the door of the refrigerator. His voice was deep, and he spoke with the practiced diction of a preacher. The family's parish had just concluded a week of Vacation Bible School, the theme of which had been "Ocean Commotion." The next day was Sunday, and Kimberly was cross-legged on the floor attaching construction-paper fish and seashells onto offering buckets for the service. "You should have given him the water from the tap," she said. "So he could taste the real Antwerp water." She emphasized the word *real*.

"The *real* Antwerp water? What do you mean?"

"The tap water has this strong sulfur taste. We can't stand it, so we use a filter."

"Where does the sulfur come from?"

Michael said, "You've probably heard about the Great Black Swamp. Because of the swamp, we sometimes get a lot of sulfur in our water. And right now we have it because when they drill a new well, they don't know whether they'll hit a sulfur spring or a fresh one."

"The sulfur is left over, then?"

"Must be."

"You know," Kimberly said, without looking up from the construction paper she was affixing to a blue pail. "If you're interested in the swamp, you might like to know that in Van Wert, at a Methodist church there, they have a sanctuary built from lumber that was tithed by the congregation while they were clearing out the forests. You might want to stop by and see it when you're done with your trip."

"You're kidding," I said. "Are there other churches like that? Sanctuaries, I mean."

"Not that I can think of—Michael, do you know of any?"

Michael said he didn't.

"I never thought of tithing with lumber," I said.

Kimberly shrugged. "Well," she said. "It's what they had."

I slept in the backyard under a clear sky, and the next morning started off again. But first I had to go inside to use the restroom and change the bandage on a blister. Before I left, I took a glass and filled it with tap water. I could smell the sulfur before it even reached my lips, a ripe waft of gas that made me wince involuntarily. Sulfur is often described as smelling like rotten eggs, but the tang was different still; it was strangely chemical, as if I were sipping rank, diluted pool water.

Walking out of Antwerp along Canal Street, I could see the impression of the old Wabash and Erie Canal in the ground (Kimberly had suggested this route to me for that reason). It was a wide, low ditch, maybe twenty feet wide and a couple of feet deep, that ran right up to the edge of the pavement, as if somebody had just pressed down hard on the earth and left an indent two hundred miles long. The canal bed, or what remained of it, was long since grassed over.

Meeting an intersection, I turned left—north—toward the river, and hit the bottom edge of Antwerp's meander loop. There I stopped to see the Maumee, following a set of leaf-laden steps down a hillside flush with trees. Thin metal rails along the steps had been painted a glaring white, some of which had dripped onto the leaves around the base of each support. At the bottom of the staircase was a wide meadow. Across the meadow, heavy, wizened trees hung out above the water—a shadow, perhaps, of the one-time forests of the Great Black Swamp. I kicked around there a little while, looking at busted-off tree branches and ripples billowing in the Maumee. The meadow was recently mown, and extended from the hillside out forty yards or so until dissolving into the river. I read some time later that this was a floodplain, that if it had been the rainy season—or, for that matter, if it had been two hundred years earlier than it was—I would have been standing underwater.

CHAPTER 4

For the next two days I walked through fields—or rather, between them. Post-swamp, the whole Maumee River Basin is basically a grid. Viewed from above, the land looks like a patchwork quilt: squares of farmland stitched together with roads, tree lines, and railroad tracks. This is by design. It's difficult to govern a region that has no clear sense of political organization, and political organization is hard to pull off in a place that is lacking in geographical organization. When the United States first acquired the area west of Pennsylvania, north of the Ohio River, and east of the Mississippi River from the British, at the Treaty of Paris in 1783, in the eyes of the new government that's exactly what it was: an enormous patch of unstructured land.

In 1784, fresh off a military victory in the American Revolution, the US Congress, under the leadership of Thomas Jefferson, passed a land ordinance that suggested dividing the newly acquired territory into states. The next year, Congress passed another ordinance that was more specific. It laid out a survey system for dividing up the land according to a Cartesian grid: a rectangular structure that segmented the country into blocks six miles square. Each block formed a township. Each township was subdivided into thirty-six one-square-mile sections of 640 acres each. And each section was then subdivided into forty-, sixty-, and one hundred and twenty-acre plots, which could be sold off to individuals. Having provided for the future of their newest real estate holding, in 1787, Congress gave it a name: the Northwest Territory.

A geographer and surveyors were appointed and deployed to mark the lines on which this new organization would be based. They began with a baseline—a primary latitudinal division— and a prime meridian, which is the same as a baseline, but runs longitudinally. All subsequent lines were drawn at mile intervals from these. The ordinance stipulated that "the lines shall be measured with a chain; shall be plainly marked by chaps on the trees and exactly described on a plat; whereon shall be noted by the surveyor, at their proper distances, all mines, salt springs, salt licks and mill seats, that shall come to his knowledge, and all water courses, mountains and other remarkable and permanent things, over and near which such lines shall pass, and also the quality of the lands." Surveyors were paid two dollars a mile to cover salary and expenses. They placed monuments at the corner of every one-mile section. Between these marks would run the earliest roads.

Errors were inevitable. The geographer and surveyors were instructed to "run and note all lines by the true meridian" and "pay the utmost attention to the variation of the magnetic needle," but there were honest mistakes. There were dishonest mistakes, too, when lazy surveyors skimped on their platting. Some sections ended up too big or small, or varied in angle from the baseline axes. Squared-off or not, this was the rudimentary outline of the Public Land Survey System, which remains the dominant means of dividing and labeling land in the United States. All existing surveys were considered canon; ultimately, any new survey would have to work from the originals, no matter how disjointed. In the process, the Northwest Territory became the first planned settlement in the country.

Back east, the former colonies had been laid out according to "metes and bounds," a survey system that used natural landmarks like ridges, rocks, trees, and streams to divide up land. With metes and bounds, an individual would first settle a piece of land, then pay a surveyor to come out and mark it. Surveyors would identify a couple of major landmarks and measure compass lines between them. Sometimes they didn't even run the lines, just marked down the rough coordinates of the property. While Thomas Jefferson argued for the rectangular grid in the Northwest Territory, Alexander Hamilton was in favor of the metes and bounds system. Hamilton believed that Americans would settle faster if they could simply move in freely, without prior subdivision.

Native tribes took a less stringent approach to land tenure. Property rights were based on occupancy and generally respected by everyone else. Often, individual plots were controlled by a family or clan, and passed on through inheritance. People clustered in villages, and together—as tribes—would occupy and control a larger area. Tribal lands were often defined by natural features, like the watershed of the Wabash River or the headwaters of the Maumee. One reason the Miami didn't want to give up their valley to the United States was that there was a connection between themselves and the land that the Miami knew better than to sever. A Miami chief, Akima Pinšiwa, English name Jean-Baptiste Richardville, said in 1832 that "Here the Great Spirit has fixed our homes. Here are our cornfields and our cabins. From this soil and these forests we derive our subsistence and here we will live and die. I repeat, we will not sell an inch of our land."

What happened next, broadly speaking, was that settlers moved into their subdivisions and the Miami were pushed out. It was more complicated than this, of course. There was a good deal of violence on both sides in the meantime. More treaties were signed, in 1834, 1838, and the Treaty of the Forks of the Wabash in 1840. The Miami, led by Chief Richardville, inexplicably agreed to a treaty that exchanged their 500,000-acre reservation in Indiana for one the same size west of the Mississippi. Some scholars say the Miami leaders may have thought everybody would still get to stay on their individual family plots. Others think Chief Richardville may have simply decided that removal was inevitable and tried to secure the best deal possible. I haven't come across anybody who thinks the Miami wanted to leave.

The Miami were supposed to be gone within five years but stayed on for six. War and disease had ravaged their numbers. Sixteen years after the passage of the Indian Removal Act of 1830, there were only about six hundred Miami in the area, and they would not leave voluntarily. So they were ousted by force; the United States government rounded the Miami up and loaded them onto boats. On October 6, 1846, removal began in Peru, Indiana. On October 7, it continued near Fort Wayne. Many of the Miami ran away or tried to hide. Some climbed to the tops of trees or hid in the swamp. White men hunted them down, "like wild animals," one eyewitness said. Some people went to the graves of their families and grabbed stones and dirt clods for the journey—fistfuls of homeland. One witness, Rose Carver, said, "They didn't want to leave their land. They just picked up a little handful of dirt and put it in a tobacco sack and take that with 'em."

Here was one use to which the canal was put during its brief lifespan: shipping the Miami in boats to Junction, Ohio, en route to Kansas. Indiana soil in their hands.

Fields ran larger, farther, nearly out of view. They appeared as huge splotches of green and gold. It was mesmerizing, almost pleasing, in the way that a painting of abstract color fields is pleasing. Walking along these roads, the cumulative effect was that of an assembly line—hundreds of anonymous blocks of land lined up one after another, the legacy of Jefferson's rectangular survey grid.

There is a great deal of interdependence in the biotic community, and what Jefferson and the Congress had done in the

1780s was to impose a standardized grid on a landscape that could not be so neatly divided. Some of the survey lines would inevitably split ponds, forests, trees, and, for that matter, the Maumee River. One parcel of land might have excellent soil, but no water, the nearby pond belonging suddenly to an adjacent property owned by someone else. A stand of trees could be split apart too, though it had grown up as one ecosystem and could not survive one part without the others. The relationship between a field and a forest, or a river and a swamp, had no bearing on how the grid was produced. At the same time, it had serious bearing on just about everything else, ecologically speaking.

Perhaps the most glaring example of the United States' lack of concern with ecological interdependence was the million-acre swamp it would wipe from the earth sixty years later. Maybe people didn't know what they were destroying. Maybe they didn't care. But it would have saved a lot of headache if they had left the swamp there. Swamps are natural flood control agents, given their capacity to absorb runoff and release it slowly. An acre of wetland can store up to 1.5 million gallons of floodwater. Wetlands also retain nutrients, pollutants, and organic wastes, filtering them out of streams. Birds and other wildlife find habitat and nesting sites in places like the Great Black Swamp. Wetlands are extraordinarily rich, biologically speaking. As ecosystems, they rank with tropical rain forests and coral reefs. It used to be that there were more than two hundred million acres of wetlands in the continental United States, but there are less than half that many now. About 60,000 acres more are lost each year.

The one-two punch of clearing forest and draining land led to the arrival of even more settlers. In nine counties of northwest Ohio, including the area formerly occupied by the Great Black Swamp, population growth was slow until the 1820s. From 1830 to 1840, with the development of the Maumee and Western Reserve Road, it jumped tenfold. Most of that growth happened outside the swamp itself. After the swamp was drained, between 1850 and 1860, population in the area doubled. By 1880, it had doubled again. The soil was unbelievably fertile. Farming is never easy, but with loam like that you could do all right. The Midwest was already integrated into a commodities market, and had been since the early 1800s. Crops from midwestern states were taken by rail to Chicago where they were stored and then shipped to the east coast.

As more of the old swampland became available, people wanted to farm greater and greater acreages. But a person, even a person with horses, can only cover so much ground, especially when planting and harvesting is involved. What they needed were machines. The Industrial Revolution delivered in the form of tractors. The first successful gasoline tractor arrived in 1892. After a couple of world wars pulled millions of horses and young men from the fields, the tractor became an indispensable piece of farm equipment. Tractors could do the work of many people, and they worked for gasoline. When the Second World War was over, many of the men never went back.

Meanwhile, farmers, on the advice of scientists, started to grow only one crop, and lots of it. The logical crops were those that could be processed and stored for a long time, like corn, wheat, and soybeans. A big field of a single plant is called a monoculture. It takes a lot of phosphorous-based fertilizers and other treatments to maintain something so unnatural over a long period of time, because the soil doesn't get a chance to replenish the nutrients that are pulled out year after year. Agricultural schools, and the companies for which their graduates worked, developed fertilizer products in laboratories that were widely applied to the fields, on top of more naturally occurring fertilizers like manure. Now, when rainwater washed across the land on its journey to the river, it took with it topsoil and both kinds of fertilizers—natural and synthetic. When people started using pesticides to keep bugs off their crops, the rainwater picked those up too.

Synthetic fertilizers and pesticides made farming extremely productive—and profitable. Decades of unprecedented output led to nicknames like "The Breadbasket of the Midwest" and "The Corn Belt." The land encompassed by that second term stretched west as far as Iowa and North and South Dakota. By the mid-1900s, the old Great Black Swamp—as well as most of the rest of the forests, wetlands, and tallgrass prairie in the middle of the country—had been overrun by acres of cropland with the genetic diversity of a parking lot.

Landscapes are always under the influence of one force or another. All along the Maumee I found the earth marked by the comings and goings of people. There were yards and homes, factories and fields, roads, bridges, railroad tracks. The results of human enterprise were present nearly everywhere I went. One of my favorite things about the Maumee River is that it does not seem to care. That much is evident in the complaints people have had about it over the years: it is too shallow and too curved; it keeps flooding towns and fields; and, for a long time, it was surrounded by the most dreadful, impassable swamp many of them had ever seen. But there it sits, stretched across the middle of the Northwest Territory, maddeningly unimpressed with what people have in mind. Even now, one mile to the next, it's hard telling how the river will behave. In this great big patchwork quilt—the cornfields and city blocks, the boundaries and baselines drawn as close as possible to square—the Maumee is a thread gone rogue.

———————

Notes from the road between Antwerp and Defiance—twenty-five miles, give or take:

A hillside full of purple flowers (pretty sure they're lavender asters). Stopped to look and saw a break in the trees. There was a path that went down to a rock beach by the river. Campfire leftovers on the beach (burnt logs, ash). Looked recent. Grassy island in the middle of the river. Thought about trying to get onto the island but it looked too soggy. Went back up to the road and kept walking.

Highway 424 became County Road 424. Not sure exactly where as the experience was roughly the same. Pavement—gravel—grass berm—fields—sometimes river coursing near. Fields are corn and soybeans, from what I can tell. They are not all exactly alike, but I have to concentrate or I might start to believe it. Most have not been harvested. Some have. Some are mostly empty with leftover stalks the color of rust. A few fields have grain bins and complicated-looking equipment around or in the middle of them. They look like factories. Some of them are covered in rust too.

Passed north of Cecil, Ohio. Cecil was also on the canal, also made charcoal from trees and sent it to Toledo. Stopped to peek at the river through a thicket of trees. Had to climb down, but the thicket was right next to the road so not far. By the water, could not see the road. Like a parallel world, timeless. Trees leaned out over the river, shading its aged face. One of them, a silver maple, looked to have at least four separate trunks. Some hollowed-out trees, with gaping holes that I could stick half my body through. How old were these trees? How old was any of this? Felt I was the only changing thing there. Stayed for a while anyway. A mile later stopped for the day.

Next day decided to stay on 424. There are clumps of flowers in the space between road and field most of the way so far. The clumps— the flowers, too—are purple (asters), yellow, and white. Brilliant moments of color in a sea of green, brown, and pavement. Picked some and slipped them into a loop on my pack.

Power lines are pretty much everywhere. They tick by slowly, regularly. If I ever got lost out here I could just follow the power lines and I would eventually come to a town, or at least a house. I passed one mailbox that was sided like a house. Next to one rusting field is an old grey clapboard barn with huge gaps in the walls. I think I could probably blow it over. Can't imagine anybody is still using it, anyway. My shirt is soaking wet. I can feel the sun hot on the skin of my arms. No clouds in the sky. So much sky. I remember a Zen saying—how does it go? Something about how the sky is the biggest thing there is.

Sat in shade by a treeline. Food: peanut butter and honey wraps, apples, nuts, Clif bars. (I should have packed more variety.) Trees: oak, silver maple, poplar. Probably more. Lots of tall grass with grasshoppers leaping around. I burn a tick off my shin with a match. Check for others and don't find any. A groundhog scuffles through underbrush. He sounds like wind among the leaves, but there is no wind.

North of Junction—where the canals met in the old days. No other people walking around. Some in cars. Two people (one compact car, one pickup) stopped and offered me a ride. I said thank you but I wanted to keep walking. The truck (black, single cab, Toyota) had a yellow sticker in the window that said "The Lord's Army."

A sign in the grass alongside 424:

> *WE HAVE GREAT WATER*
> *DO YOU?*
> *800.866.5585*

I saw a lot of signs during my walk. Some of them homemade. I like signs because I like to know what people think is important enough to make sure other people read it. Some of my favorites: a spray-painted, plywood roadside sign advertising fifty-five gallon barrels for $10 each; the words "Suit Up" in black script on a white barn (no mention of when or what for); just before Antwerp, a church sign pun, "The sun may burn but the Son will heal." Just east of Fort Wayne, I had passed a sign next to an enormous cornfield that read:

BRODBECK SEEDS
It's Personal!

Three miles and zero human beings later, another:

BRODBECK SEEDS
It's Personal!

It was late afternoon when I walked out of the fields and into Defiance, Ohio, four days after leaving the confluence. The first restaurant I came to was Jigg's Drive-In, a fifties-era leftover with half a dozen cars pulled up under the awnings. The building itself was small, with three walls of windows. The interior had orange walls and a white corrugated ceiling. There was a door on either side, and both were open. Between the doors stretched a worn wood counter. On the right side of this counter, three servers leaned with their backs to me, under a sign that read: "Waitresses Only." So I ordered on the left—two chili dogs and a root beer that came out in a glass mug with about an inch and a half of froth— under a sign that read: "Place Carry Out Orders Here." Then I sat down by a row of windows across the room from the counter. I ate quickly, but lingered a while to rest my feet.

I had booked a room for the evening at the Elliot Rose Guesthouse, a bed and breakfast to the south of downtown. The guesthouse is a pleasant-looking white structure with a flat front profile, bay windows jutting out on the left, and flowers growing on a fence out back. It is owned by an older couple, Paul and

Diana Bauer. The place was named for their granddaughter, Elliot Rose. Paul met me at the door wearing a red Ohio State Buckeyes sweatshirt. "We're usually around all the time," he told me, "unless the Buckeyes are playing." My room was on the second floor with a shared bathroom. There was an upstairs kitchen with cereal and bananas in case I wanted a snack. Milk was in the refrigerator and tea on the counter. Since I had already eaten dinner, Paul recommended a place downtown that had excellent pie.

I cleaned up a bit and went downtown with my camera. It was a Monday night and most of the stores were closed. One storefront caught my eye because it had a large sign that said "HEROIN" in red hand-painted lettering four feet tall. I looked closer and saw that it also said "KILLS." "KILLS" was only about eight inches tall and it was written vertically to the right of "HEROIN." A few streets over, I found the place where the Auglaize River joins the Maumee. This was the location of Fort Defiance, one of the forts Wayne built charging down the river in 1794. To commemorate Fort Defiance, the city put in a park overlooking the joining of the two rivers. The park has a grass lawn, a tall flagpole with an American flag, and two full-size iron cannons aimed out toward the confluence, one on either side of the pole. Some families were picnicking in the park, and teenagers were holding hands and kissing.

Curving around the front edge and just below the park was a riverwalk. I went down there and was arrested by what I saw: a layer of bright green algae slicked across the surface of the Maumee River. This was the first time I had seen an algae bloom, outside of pictures. Truthfully, it looked more like an oil spill than anything. The disappearing sun lit the algae from the side and gave it an unearthly, neon glow, like something out of a cartoon. I took a set of steps down to a small, gritty beach with tufts of grass, sticks, a black plastic trash bag, a Styrofoam coffee cup, and an Aquafina bottle full of algae. Above me, two white-haired women in athletic clothing carrying colorful one-pound weights paused their walk to lean on the rail. One of the women said, "Disgusting." The other woman said, "Yes, isn't it a shame." On the water, a boater and passenger motored through. I stood around for a minute and went back up through the park. Nobody there seemed to have noticed the phosphorescent river. They went on picnicking and necking as if it were the most normal thing in the world.

Fort Defiance was not the only thing to happen here, present activities excluded. A small plaque buried in the ground at the edge of the park said that some buffalo were seen across the river from this spot in 1718, that there were trading posts here in the middle of that century, that the first permanent US settlers arrived in 1817, and that the last of its Native residents were removed during the 1830s. Next door to the park, at the Defiance County Public Library, I found a small room in the back dedicated to local history. The shelves were filled with items of local interest—newspapers, genealogies, school yearbooks. Some of the documents were in plastic sleeves. On the walls hung maps and other framed items. In the middle of the room was a desk with a large book on it. The book was an illustrated history of Defiance County from 1883. It began, "A hundred years carries us back to Ohio in a state of nature; its forests unbroken by the labors of civilized man; its rich mines unopened; its beautiful lakes and rivers free of all navigation save the Indian canoe."

It was a familiar idea then: that the natural world was simply raw material for the pursuits of "civilized man." Such vast wilderness, such unbroken expanses of forest, seemed boundless, and therefore inexhaustible. The wilder regions were considered obstacles to be overcome; certainly the Black Swamp was antagonistic to settlement. It was to the white man's credit that he tamed this defiant wilderness and made it profitable. By the mid-1840s, the territory had been marked off, chopped up, and sliced through. Native tribes had been kicked out or suppressed. A road split the Great Black Swamp, and men were mining timber like gold. A hundred years later, the swamp was gone, the basin revolutionized into an enormous, fertile field, diced up and sold off. It's basically still like that—the whole watershed, farms and cities alike, divided and subdivided to the point that it is somehow possible to imagine each little box exists independent of all the others, and of the river too.

Passing back through downtown on my return to the Elliot Rose Guesthouse, I made a detour toward a slice of pie. But I had waited too long—the diner, like the library, closed at eight. I settled for a large bowl of cereal in my room. The guesthouse had a library, where I found a book called *Maumee River 1835*. It was built around the journals of William C. Holgate of Utica, New York. William C. Holgate had traveled with his father to Huntington, Indiana, that year, including a stint up the Maumee from mouth to head. Huntington sits at the Forks of the Wabash, near the home of Akima Pinšiwa. Flipping to the entries around Defiance, I learned that the Holgates had arrived in the early part of June. "There are but few buildings about Defiance," William Holgate wrote. His father, who was on an expedition to buy land for development, bought five village lots and an eighty-acre plat across the river for $2,500. Upon leaving Defiance, the Holgates went twenty-two miles upriver through dense swamp, by "hard tugging through the mud," and spent a night near modern-day Antwerp. This was the stretch of river I had only just completed.

The next morning I packed my bag and ate breakfast downstairs. My friend Jason Bleijerveld was on his way from Indiana to pick me up for two days of canoeing. I was looking forward to letting my arms have a turn. And it would be nice to spend some time on the water, rather than adjacent to it.

I was the only guest at the table. Diana sat across from me. Elliot Rose played games on an iPad at the other end, and Paul puttered around the kitchen. I ate spinach frittata, fruit, and orange scones, and talked with Diana. Diana had silver hair that was longer on top than on the sides. She wore dark-rimmed glasses, and a navy blue crew neck with a polo underneath. Behind the glasses her eyes were soft. In the course of conversation, I learned that Diana's father had a farm along 424 just outside of Defiance, which he still lived on. "You probably walked right past it on your way in," she said. "The front half is still farm, but the back part, on the river, he's set back to trees." I asked what kind, and she said: "Black walnut, sweet gum, honey locust, white pine, and then there's one other kind of evergreen. The back part there floods when it rains in the spring, which was good for the crops. That river bottom—it's so fertile you could just stick your hands in it, it's so rich."

"Hold on," I said. "Let me get my notebook."

Diana's family had come north around 1840 on the canal. They were German immigrants who arrived in America and then went wherever there was work, which in this case meant following the canal as far as necessary. "The culture of the time was that you farmed, and you had enough farmland to divide it up among your sons," she told me. "And my great-great-grandfather had eight sons so he had to come far enough north that he could have enough land to give them!"

I copied this down. Paul came out and leaned on the chair behind Diana. We started talking about the guesthouse, and I learned they had bought it a few years back to renovate themselves. They liked working on houses. Their personal home had been an ongoing project for most of their marriage. Diana said, "Our philosophy is that we'll take a place and clean it up, take as good a care of it as we can, and then we'll eventually pass it on to somebody else who will hopefully do the same thing we did. That way these old houses can keep on living. There's just something there that you can't find in new houses. There are a lot of young people around here buying houses now and fixing them up. We love to see that. With this house, I've wanted to see what it looked like originally. There was an old picture of this street out here from around 1910, but the trees are all about ten or twenty feet high so you can't see any of the houses because they're in the way. But the whole street looks totally different now; it's amazing what a century can do."

CHAPTER 5

We—Jason Bleijerveld and I—put in our canoe downstream of Defiance at Independence Dam State Park. Just past the dam, on the north bank, was a parking lot with cement stairs angling down to the river, which we navigated with some difficulty. To our left, the lower Maumee was shimmering in the mid-morning sun, tossing the light back in ever shifting arrangements so that each moment reflected some new contour, some new dimension. The canoe was loaded in the middle with a few lidded buckets of food and gear, and a dry bag of Jason's that resembled a black nylon tube sock, rolled up at one end and watertight. With our weight in front and back, the canoe rode too low for the shoals of our launch site. We decided to walk out a hundred yards toward what looked like deeper water.

It was uneven going. One moment the water would only lick my ankles, the next I would be in above the knees. Horseflies buzzed about our heads and arms. A fish rotted on a flat stone in the baking sun. The river was shallow and unreasonably warm and stagnant, like tepid bathwater. We could only just make out the bottom. We were not confident in our steps, given the murkiness of the water, and so made them slowly and deliberately. I managed to keep my balance for the most part, but half an hour in I slipped hard, forward, and bashed my knee on a rock. I sprang up quickly to see a gash near my kneecap. Blood and water mingled. It was an uncomfortable sight. Whatever was in that water was now, possibly, in me (and vice versa). I winced when I thought of the dead fish and the algae I had seen upstream.

What we thought to be the end of the shoals was not. Instead, we slogged for the better part of an hour before reaching water that could support us. We began to paddle, Jason in back and me up front. It didn't take long to find a rhythm. Jason is a more experienced canoer than myself, having run dozens of rivers across the country, including many faster and more exciting than this—not to mention clearer. For most of the morning I could not see much past the throat of my paddle. I said that this opacity added to the river's mystique, and Jason reminded me that a river will tell you plenty about itself if you are paying attention.

The first thing a novice canoer learns to read is the current: which direction it's flowing and how quickly. Then comes turbulence. The surface of the river changes in response to what is underneath. Rapids are the most obvious example, but not the only one. Whenever water moves over some sort of obstruction, its flow adjusts: a 'v' pointed upstream means a submerged rock or log or stick; a line of turbulence means a row of rocks, or a possible ledge; pillowing water means something is sticking through. If the water is circling, that's a hydraulic move that means there's a break in the flow somewhere, a hole created by some change in topography. On the other hand, a flat, glassy surface, especially when there is little to no wind, usually signals a deep spot. A river is constantly betraying its secrets.

One thing that is not a secret, though it still sounds wrong to say it out loud, is that the Maumee River is not made up exclusively of water. The Maumee is an alluvial river, which means its bed and banks are for the most part lined with soil or sediment that is constantly dredged up by the current and carried downstream. This sediment is suspended in the water along with anything else that happens to enter the stream—including "floatables" like leaves, sticks, cigarette butts, and motor oil. When you look at the river, you're staring at a lot of different things all jumbled up together. A certain amount of suspended sediment is natural in alluvial rivers. But the Maumee receives a great deal extra from outside its banks. Much of the land around the river is naturally poorly drained because its clay soils do not absorb water well— hence its former life as a swamp. But because of the various efforts of people to force drainage anyway, most of the rainfall rinses off into the waterways, bringing with it topsoil from across the basin. The Maumee River has probably never been crystal clear, but it carries a lot more sediment since it lost the Great Black Swamp.

Jason and I spent most of the morning out near the middle of the river where its surface was largely uneventful. From where I sat in the prow of the canoe the water stretched smooth from one bank to the other, like a sheet tucked in tight around the edges. I leaned over the side of the canoe looking for fish, but all I saw was suspended sediment. I was probably also looking at nitrogen and phosphorous. There is a lot of nitrogen and phosphorous in the Maumee River. Most of it comes in hanging onto soil particles washing off of fields. This is called sheet erosion. When rivulets of water carve small channels into the ground, the erosion that occurs is called rill erosion. When rill erosion gets out of hand, and starts cutting larger channels, it's called gully erosion. The development of a more complete railroad system in the last half of the nineteenth century meant that rivers were no longer as necessary for trade, but they have continued with the work of transportation, picking up the unabsorbed water running off the fields—water bringing with it topsoil, and the pesticides, fertilizers, and other byproducts of high-yield farming methods.

Topsoil is a renewable resource, but a slow growing one. It takes up to five hundred years to produce one inch. Most crops need at least six inches of good topsoil to grow. Nobody has three thousand years to sit around waiting for topsoil to form, so the best thing to do, one imagines, would be to conserve and enrich what already exists. This is the sort of wisdom that is lost on machines. Industrial agriculture, sometimes called conventional agriculture, accelerated erosion for most of the twentieth century. In 1939, two soil conservationists named G.V. Jacks and R.O. Whyte claimed that "more soil was lost from the world between 1914 and 1934 than in the whole of previous history." They were wrong, but even contemporary scientists acknowledge that they captured a bit of truth about the erosion of American farmland during that period: it was a long, slow avalanche, and it was gathering speed.

Erosion is not a new phenomenon. It has happened in one form or another as long as there has been soil to erode. But humans have made it worse, mainly through agriculture. One of the biggest culprits was the plow, which cut deep into the layers of topsoil and loosened it up for the disruptive forces of water and wind. Even so, for a long time, agriculture was still a relative non-issue for erosion. People simply weren't depleting the landscape fast enough to cause problems. That is, until the Industrial Revolution gave them the tools. In the US, there was a bit of a slowdown once people realized what was going on—which happened on a national scale sometime in the 1960s—but not much. By the year 2000, US croplands still averaged about four tons of soil loss per acre every year. It's not far from that now. This works out to about one millimeter over the course of twelve months, which doesn't sound like a lot, except that it takes only twenty-six years to lose an inch of topsoil this way. Even by the most conservative estimate, topsoil is washing away more than ten times faster than it is being replaced.

If the first half of the Maumee is slight and meandering, the second half is expansive and single-minded. Post-Defiance, the river is much wider than among its earlier meander loops—around a thousand feet at points—and also straighter. This is because its course here is defined largely by bedrock. The result is a broad, deep body of water moving deliberately and relatively efficiently—albeit slowly—downstream. With the exception of our hourlong stroll through the shallows, we had estimated we would be able to move at a steady clip of four or five miles per hour. Not so, said the river. It didn't help that the water's incline from Defiance to Toledo averages only 1.2 feet per mile—basically flat. It was effectively like paddling across a giant, placid lake with only two shores.

To measure our movement, I would pick a landmark on the shoreline and watch as it moved closer, then beside us, and then behind. At one point I picked out a couple of fishermen on the bank ahead, a man and a young boy. Fishing is a popular thing to do on the Maumee River, especially this section. There are about 114 species of fish in the Maumee. The most common fish caught include walleye, white bass, catfish, and crappie (pronounced "croppy"). Before the river became polluted with pesticide, there were at least a dozen more, including gilt darter, western banded killifish, blacknose and popeye shiners, sauger, several species of sturgeon, and the harelip sucker, which is now extinct.

The largest fish in the river are flathead catfish. The biggest flathead anybody has ever caught was 123 pounds, or roughly the weight of a slim high school freshman. In 1979, somebody caught one in Ohio that was 76.5 pounds. This still stands as the state record. I don't know the size of the biggest flathead ever pulled from the Maumee, but I've read that they can push fifty. Flathead stick to covered areas of the river and are usually most active in the dark. Fishermen looking to land the big ones show up around sunset and stay until after midnight. They bait their hooks with small live bait, like bluegill or goldfish. Most of them don't eat their catch, but toss it back into the river. The Ohio EPA recommends not eating more than one meal a month of flathead catfish from the Maumee River because of the high concentration of mercury and PCBs in the water. For channel catfish, it's once every two months.

PCBs are a group of man-made chemicals that were manufactured in the United States beginning in 1929. They were used in everything from plastics to thermal insulation to transformers to motor oils. PCBs can take the form of a liquid or a solid. They are useful in the industrial world because they are chemically stable and not very flammable. They are also toxic. The Toxic Substances Control Act of 1976 gave the EPA control over the production and distribution of certain dangerous chemicals, including PCBs, asbestos, radon, and lead-based paint. After 1979, nobody was allowed to produce PCBs commercially in the US. But, because they are not a naturally-occurring substance, PCBs do not break down. They are easily transported by water, and people have found them all over the world, even in places far away from where they would have been released. By 1979, enough had already been created to cause problems thirty years later.

Most experts think the Maumee reached its water quality nadir sometime in the seventies. Since then, the river has seen steady improvements in nutrient and chemical loads. Most people agree that it's clearer now than it was in the 70s. It used to be that people who went swimming for more than half an hour would come out with their fingernails stained yellow. That doesn't happen anymore. This was some comfort, though I had not stopped thinking about the river water seeping into the hole in my knee. I thought when we stopped next I should buy some peroxide to sterilize the wound.

It took five hours of steady pulling to reach the city of Napoleon, a trip of fourteen miles. We tied the canoe to one of a series of narrow piers at a park upstream of the city. The only shade at this park came from a restroom structure. We were sweating profusely, having spent the past few hours baking in the sun which burns somehow hotter in the middle of the river. We filled our bottles from a water fountain and sat down in the shadow of the can. It was too hot to be hungry, but we ate a late lunch anyway, warm vegetable sandwiches and leftover scones that Diana had wrapped up for me that morning. After lunch we decided to walk into Napoleon to buy ice cream and peroxide. It was about a mile walk to town, further to the first place that sold ice cream. By the time we made it there and back we'd lost an hour and a half of time.

Meanwhile, clouds had grown over the sky. A chill wind whipped upstream as we paddled down, sliding under an enormous bridge that connected the two halves of the city. Napoleon came right up to the river here on both sides, its sloping borders studded with large rocks and ragged wedges of cement. From the river, downtown Napoleon smelled distinctly like manure, though it was unclear from where this scent came. We had not seen any cows. We did see a couple of pipes jutting out over the water. There was a Campbell's Soup plant down from the bridge, and the manure smell gave way to the stale, overwhelming odor of canned chicken and noodles. A turret in the shape of a water tower, plastered with the Campbell's logo, rose out of the plant, which hummed intensely the monotonous music of industry. It took a good half hour of hard paddling for us to leave its song behind.

Downriver of Napoleon, the city gave way to a thin line of trees, their mangled roots jutting out from the banks like a ribcage. Eventually, even these were gone, and the spare-cut shore stood high and sharp. Rows of corn stalks marched above. Beyond them was the sky, which curved away out of sight behind the stalks. It gave the illusion that the world continued on forever like this: only corn and sky. We soon came to a stretch of river that coursed through heavy forest on both sides. The clouded sky began to pull itself apart, revealing the sun. Looking for shade, we moved out of the middle of the river and closer to the banks. These were not nearly as steep as they had been back toward Napoleon, and were noticeably less eroded. Jason remarked that the water here seemed cleaner. I could see more of my paddle.

It turns out that one effective way to cut erosion is to have a lot of healthy forest and prairie cover around, and a wetland at the water's edge. But this is not the only way. Maybe the best way is to keep soil from washing off fields in the first place. In the 1980s, due in part to incentives in the 1985 Farm Bill, many farmers began to practice conservation tillage, or a variant called no-till farming. In both practices, farmers leave at least 30 percent of crop residue on the surface of the field, which protects the soil from wind and water; cropland erosion dropped more than 40 percent between 1982 and 1997, mainly because of conservation tillage. There are also some less intuitive approaches that have been tried. We paddled another half dozen miles before I realized the jagged cement blocks in Napoleon were dropped there to help stop the bank from eroding away.

I was becoming tired of dragging myself through water. Jason was too. Our arms and shoulders burned, then stung, and eventually—mercilessly, I thought—went almost completely numb. My right shoulder had begun to make a popping noise, and my wrists were tingling. Exposed skin was red with sun. Jason and I paddled on, alternating sides. As the afternoon wore into evening we began to switch more frequently to give our bodies a break. By mile twenty we were shabby.

Our ambition had gotten us into a long intermediate stretch of river in which there was no place to stop. On either side we faced expanses of state forest, interrupted only occasionally by a bridge, or by a mansion and flawlessly manicured lawn. Neither of these was the kind of place you could pull up a canoe and camp for the night. We didn't feel like trespassing, and even if we had wanted to, the forest was thick as jungle all the way out to the edge of the water. Exhaustion set in. Even talking felt like a chore.

We did not stop out of fear that our arms might take the opportunity to go on strike. Three more miles. Then two more. The sun began to set, though I do not remember when or what it looked like. In my memory there is only forward, an unbroken ribbon of water wrapping out ahead.

At last, we came around a bend and saw a cluster of RVs on the southern bank. I had never before found an RV park to be beautiful, but I had also never before canoed twenty-five miles against current-less water. The trip had taken us more than eleven hours. Jason said it was the worst he had ever felt after a day of paddling. Even my bones were tired.

There was a small boat launch near one end. As we pulled closer, I could see that all the lots along the river were full.

"Uh oh," I said.

Jason said, "I will sleep on this ramp if I have to."

We ran the canoe up the ramp, got out, and pulled the boat far enough out of the water that it would not slide back in. Then we went looking for an open lot. The park consisted of rows of campsites separated by long gravel paths. Next to the park were cornfields. Every single spot was taken. I learned that people rent out sites for an entire summer and leave their RVs there. It is a whole lifestyle. Surrounding boats and RVs were sets of outdoor furniture, tables, and umbrellas. Flagpoles and flags. Some people had built wooden decks that looked to be more permanent than the actual dwelling, and probably more structurally sound. One person had even put up a trellis. But the park was basically deserted. We could not find an office. It was a Tuesday, so everybody must have been at their homes elsewhere. Most of the RVs were dark and drowsily quiet. There were a few holdovers: down one row we could see a fire going, and a ring of people. One shirtless, large-bellied man drove slowly up and down the rows in a golf cart, blaring Whitesnake from a handheld radio. He stared at us as he passed.

Jason and I wandered the gravel paths, and considered squatting next to one of the empty trailers. The only open grass we found was on the end cap of one of the rows, a patch of rough ground that had been left open, presumably because it was too small to fit an RV. In the middle was an enormous halogen security light. The spot was at least a hundred yards from the boat ramp. My shoulders hurt just thinking about carrying the canoe that far. But neither Jason nor I felt good about leaving it on the ramp all night. So we hauled the gear a bit at a time: buckets one by one, my hiking pack, then the dry bag and paddles, and finally the canoe—the empty hull, in fifty-foot increments, grunting like tennis pros with every step.

By the time we got the tent set up it was after ten o'clock and dark. Too tired to prepare a fire, much less a decent meal, we ate apples, handfuls of trail mix, and peanut butter straight out of the container. It was inappropriately hot. We slept shirtless without bags and left the rainfly off the tent for airflow, though there was none. I tried to lay on my sides, then my back. Every position hurt my shoulders. Jason muttered something about a sunburn. The halogen light blazed overhead. I wrapped my shirt around my eyes and sweated myself to sleep.

We woke the next morning with the sun. Half an hour later we were on the water. The river stood flat as glass, and we slipped through it quietly, the only sound that of our paddles dipping in rhythm. Much to our surprise, neither of us felt too deeply the effects of the previous day's rowing. Much more uncomfortable were our stinging sunburns. To avoid making these worse we had both replaced our shorts with lightweight pants, and put on loose long-sleeved shirts. A male great blue heron traveled just ahead of us down the river. He moved in intervals, a few hundred yards at a fly.

Between flights, the heron perched among the trees that grew near the edge of the water. Each time we approached his parallel, he would take off again downriver, landing just at the edge of sight. I did not have out my bird guide, but a blue heron is easy enough to identify.

Here is a partial list of the birds we saw in our two days on the water:

Kingfisher (maybe 8, making about 10-12 for the trip)
Great blue heron (maybe 10 while canoeing)
Green heron (1)
Some crows (not many)
Geese (of course, clustered near popular areas)
Bald eagles (3 close together, plus fledglings—must have been a nest)
Gulls (too many to count, especially around the Grand Rapids dam)
Jason saw two swallowtails, and both times I thought he meant birds when he actually meant butterflies.

We trailed the heron all morning to Grand Rapids, where we had to portage around a dam. Following a shaded channel, we came across a set of docks near downtown and got out to stretch our legs. It was late morning, maybe eleven o'clock. We hadn't eaten much of a breakfast, so I suggested we wander into town and get something to eat. Making our way down Front Street, we found a restaurant called Miss Lily's, sat down, and ate two big bowls of breakfast scramble. We took our time on the way back to the boat. When it comes to making rivers accessible to people, Grand Rapids has it figured out. The channel up which we had paddled was separated from the river by a long earthen island. On top of the island was a park. Grand Rapids Park was connected to the town by bridges. There were benches in the park, and pavilions, where you could hang around and watch the river rushing over rapids.

Everybody we talked to in Grand Rapids was pleasant; nobody mentioned that the town marks the beginning of seven or so miles of unreliable depth on the Maumee. When we re-entered the river just south of town, our canoe encountered a ledge of submerged rock almost immediately, and ran aground. It was the first of many such instances that afternoon, the riverbed consisting mainly of chunks of rough stone, crookedly aligned, rising suddenly and frequently to the surface. Some of the rocky stretches were labeled as rapids—the Maumee River Rapids and the Bear Rapids, for example—which might lead you to think that they are fast-paced and exhilarating. You would be wrong. They are, however, extensive. The rest of the day we spent in and out of the water, walking a couple of hours total through shallow, riffling current, dragging our canoe like a large, unruly dog.

Jason and I weren't the only ones to ever walk down the middle of the Maumee River. Father Joseph Pierre de Bonnecamps, Pierre-Joseph de Celeron, and the rest of their party, did the same thing in 1749. The trip is recounted in de Bonnecamps's journal, *Account of the Voyage on the Beautiful River Made in 1749, Under the Direction of Monsieur de Celeron, by Father Bonnecamps.* After recounting a disastrous trip up *Riviere a la Roche*, de Bonnecamps wrote: "The Miami River caused us no less embarrassment than Riviere a la Roche had done. At almost every instant we were stopped by beds of flat stones, over which it was necessary to drag our pirogues by main force. I will say, however, that at intervals were found beautiful reaches of smooth water, but these were few and short." It certainly was beautiful. Surrounded, as we were, by woods of such rich green, and by the water which appeared here remarkably clean, it was possible to believe that, in the two hundred and sixty years since, nothing had really changed.

One mainstream position on conservation holds that we ought to preserve the spectacular places of the world in a state of virgin wilderness. There is a powerful logic to this argument.

Who wants to lose the Grand Canyon? Who wants to see Yosemite turned into subdivisions? These places have a certain aesthetic pleasure, a picturesque beauty that makes this line of thinking attractive. The natural extension of such ideas is the formation of a national parks system and wildlife refuge plots, which are difficult to argue against. One can imagine the horrors that might have been done to that land had it not been federally protected.

But what about the unspectacular places? The no-longer-wilderness? What about the places that perhaps were spectacular once but have already been sullied? A lot of this country is not picturesque in the conventional sense, and a good deal of it has already been chewed up and spit back out. We can imagine the horrors from which the national parks have been spared because they've occurred just about everywhere else. Comparatively little of the country—comparatively little of the world—is protected by the sort of legislation that preserves Yosemite. It seems unlikely that the cities and croplands of Ohio ever will be. One reason for this is that people have to live somewhere, and so not every area can be protected in the same way. Another reason is that we simply haven't been as concerned, historically, with the health of, for example, Wood County, Ohio. There ought to be a way of thinking about conservation that includes places that don't look like the national parks. We ought to care about what is happening in and around the Maumee River.

There have been some attempts to protect rivers and the wildlife that inhabits them. For example, the federal government designates some rivers as "Wild and Scenic." The National Wild and Scenic Rivers Act of 1968 declared that "certain selected rivers of the Nation which, with their immediate environments, possess outstandingly remarkable scenic, recreational, geologic, fish and wildlife, historic, cultural or other similar values, shall be preserved in free-flowing condition, and that their immediate environments shall be protected for the benefit and enjoyment of present and future generations." The National Wild and Scenic Rivers system covers less than a quarter of one percent of the rivers in the United States. In Ohio, there are almost thirty thousand miles of rivers. Only 212.9 of them are designated as Wild and Scenic.

The Maumee River is not a Wild and Scenic River. It is a Scenic and Recreational River. Or rather, some of it is. According to the Ohio Department of Natural Resources, "The Scenic portion of the Maumee River originates at the Ohio-Indiana state line and extends 43 miles to the U.S. 24 bridge, west of Defiance." After that, it becomes "recreational" for fifty-three miles. The difference has to do with how easy it is to access. Ultimately, it doesn't really matter. The act does say that federal, state, local, and tribal governments can't support anything that will disrupt the "free-flow" of the river, or its quality (storm water overflow systems notwithstanding). But not all the land along the river is publicly controlled, and the government doesn't tell private property owners what to do. Instead, it relies on "voluntary stewardship" by landowners and river users. I am reminded how illogical it is to draw boundaries on a river.

Around three o'clock, Jason and I came to another long stretch of shallow water—our third or fourth of the afternoon—and got out to walk, shoes and pantlegs still wet from the previous stint. It took a while to find any sort of rhythm, sloshing one foot at a time over the inconsistent rocks. Eventually I called a pause, lifted the camera out of one of the tubs in the bottom of the canoe, aimed the lens toward the shoreline to my left. I framed up a shot but lost it when Jason shouted at me to look alive. I turned around to see that the rear of the canoe, caught in the current, had begun to swing wide to the right. It was tacking perpendicular to the flow of the river, which is the first step in tipping over and dumping out our gear. So I lunged to grab at the gunwale. I caught it all right, but lost my balance and ended up sitting right down in the water, canoe in one hand and camera in the other.

I think anybody who shrugs his shoulders about water quality in the Maumee River should have to go out and sit down in the middle of it. This is the best way I know of to get a person to care. The truth is that we are always this connected to the river, but no one really thinks about it like that. We have built channels and embankments and water treatment plants. We have dug drains and buried sewers. We have joked about fish with three eyes and told teenagers to stay out of the water, as if water were something that could be avoided. In other words, we have convinced ourselves that the condition of the river has nothing at all to do with us. But now I had sat down in the middle of it, and suddenly I believed that it had very much to do with me.

Jason rushed to stabilize the boat, splashing water high as he moved. We returned the camera to its bucket, gyrated the canoe back in line with the current, and carried on with me holding tightly to the stern. Half a mile or so later we reached deeper waters and got back into the canoe, sopping wet. Jason told me he was thinking about writing a poem called "Baptize Me in the Maumee River," which was to be a manifesto on living in right relationship with the waterways around us. I liked this idea, but said the only honest thing to do in that case would be to let me dunk him then and there. Jason said it was more of a metaphorical poem and that it was anyway too deep where we were, but that given the seemingly improved condition of the water, he might consider going under when we got closer to shore. Depending.

———————————

People don't get baptized in rivers as often as they used to. For a long time, it was common practice among Christian churches to perform adult baptism in rivers near where a congregation lived. Now, a lot of churches have built small baptismals into their sanctuaries. I have heard of people being baptized in pools and ponds and lakes. I don't know of any churches that baptize people in the Maumee River. I think such a practice could have merit beyond its salvific aspects. This could be perceived as irreverent or blasphemous, though I do not mean it that way. What I mean is that it would be difficult to ignore what has been happening to the Maumee River if we recognized it as sacred.

It's not a new idea. Most major religious traditions include a belief in the sacredness of the natural world, though some of their contemporary iterations have sold this down the river, so to speak. The Christian tradition is the one most familiar to me. In the Bible, as in the Miami origin story, everything begins with water: "In the beginning when God created the heavens and the earth, the earth was a formless void and darkness covered the face of the deep, while a wind from God swept over the face of the waters." And water is among the main orthodox religious symbols. In most Christian churches, baptism is considered to be a symbolic representation of a person's entrance into the community of faith. Rivers themselves have routinely been consecrated for use as a baptismal; even Jesus was baptized in a river. Some Christian churches have a specific term for water that has been blessed for baptism: holy water.

As far as I'm concerned, water is sacred. It is basic and life-giving. One of the most irreverent things I can imagine is to dump waste into a river. Water is a part of the broader community of life in which we exist, and participates fundamentally in the continuance of all human and non-human life. And the health of these waterways, even those that flow through ordinary places, that are not scenic, is a matter of some immeasurable importance. It could not possibly be a bad thing to give them more consideration than they have evidently received thus far. Holy, according to one definition I have read, can mean dedicated or consecrated for religious use, as unto God. It can also mean simply deserving of deep respect, awe, or reverence. If any earthly thing is holy, in whatever sense you'd like to take that word, certainly rivers are one.

We pulled the canoe out of the water at Farnsworth Metropark, on the north bank of the river, near the campground where I was to spend the night. Jason's father met us, and we all three ate dinner in nearby Waterville. Then we tied the canoe onto the roof of Jason's father's Subaru and they drove back upstream. I returned to my campsite, which sat basically right on the water, hemmed in by trees. Straight ahead, in the middle of the Maumee, was an island, a wildlife sanctuary covered in dense forest. A kayaker drifted through the setting sun. I built a small fire and sat down to write in my notebook. Staring out over the dark water, it seemed to expand forever in either direction. That night, I lay awake in the summer heat with the rain fly unzipped. The last thing I saw before I fell asleep was the river.

CHAPTER 6

Rivers have a natural disposition to movement. It's one of their more conspicuous features. Look at nearly any point on a river, and probably the first thing you'll notice is that the water comes from one direction and goes off in another. But where does it go—all that water, sediment, chemical, and fertilizer, tumbling around the Maumee? In the morning, I sat on the shoreline rocks, eating an apple, and wondered about that question. The obvious answer, and the one I was turning over in my mind, is charmingly vague: downstream, of course. Which is as much as to say, "somewhere else." I used to believe that. Now, I wasn't so sure. I finished my apple, packed up the site, and took off, again, downstream.

I was on the trail by eight o'clock. It was the final stretch of the Towpath Trail, which follows the route of the old Wabash and Erie Canal toward the town of Waterville. Six days into the trip, I had left behind the rural, field-heavy country—from here on, it would be city and trail. I was feeling good; two days in a canoe had given my legs a rest. By mid-morning, I would reach Waterville. By late afternoon, I would be on the outskirts of Toledo. Near the end of the Towpath trail, around 9:00 a.m., I took a flight of stone steps to the right and past a grove of trees. It led down to a low-lying stretch of ground next to the river. A picnic park.

Coming into the clearing I saw, beyond the wooden tables and rusted iron grills, a multiple-arch bridge extending across the Maumee. One of the pillars was different than the others, large and asymmetrical. It looked like an island. Back at the top of the steps I found an informational board that said the pillar was in fact a mass of limestone protruding from the ground, sometimes called *Roche de Bout*, but more commonly *Roche de Boeuf*—Buffalo Rock, presumably because it resembled in some way a buffalo.

The whole Maumee River basin is built on a layer of Paleozoic rock, limestone included, buried anywhere from 450 feet down to just a few feet beneath the surface. Some places, especially along streams, this bedrock is exposed. One such spot is at the Bowling Green Fault, which trends north one hundred miles from near Findlay, Ohio, into southern Michigan, crossing the river within sight of *Roche de Boeuf*. It is a geological rift formed in the usual way—a build-up of pressure, followed by sudden, violent displacement. When the water level is low, it is possible to see the rupture.

Roche de Boeuf was a navigational landmark in the days of European exploration, and no doubt before. The Ottawa tribe, who controlled the lands around the rivers here until the European invasion, held councils on the rock. The French made it a settlement. General Anthony Wayne, shooting his way up and down the Maumee for the United States in 1794, put an outpost on the western shore close by. Fort Deposit was a stockade consisting mainly of palisade walls—sharpened logs jammed into the ground point-side up. It was built hastily, over one or two days in mid-August, to store the marching gear of US troops. They had attack on the mind and needed to travel light.

On August 20, 1794, General Wayne advanced his army downriver from Fort Deposit toward a string of bluffs where Native forces were camped. This was before Fort Wayne, before the army's final push west, before the Treaty of Greenville. The army marched in five columns, two hundred yards apart. On either side of these, two hundred yards away, were wing columns. The Maumee, General James Wilkinson observed, was "bubbling over a rocky sheet." The day's march would be a short one. It would conclude five miles downstream, with the Battle of Fallen Timbers.

In August of 2016 I traced the same path, from the Towpath Trail onto South River Road with its enormous homes and long, shallow lawns that extend all the way down to the river's edge; through Waterville, a canal town where I stopped for a plate of scrambled eggs; along North River Road, where I caught between houses a view of the Maumee Rapids; past a manufacturing plant for commercial roofing and insulation; past a wastewater treatment plant; then, crossing north up a narrow side street, onto a road called Fallen Timbers Lane. This was the way to the Fallen Timbers State Memorial. Across the street and a mile or so north of the battlefield are a shopping mall, the Shops at Fallen Timbers, and a golf course, Fallen Timbers Fairways. The best way to get to the battlefield, by car, would be to come up Highway 24 and get off at Exit 67, Fallen Timbers Lane ("with access to the Shops at Fallen Timbers"). The stretch of 24 that runs from Waterville to the Fallen Timbers battlefield has a second name, Anthony Wayne Trail.

There were only two cars in the parking lot of the Fallen Timbers State Memorial when I arrived. I had come around to the main entrance, which consisted of a parking lot, a map under plexiglass, and a pair of pillars. A path leads from the parking lot to a tall bluff. At the end of the path, near the edge of the bluff, is a limestone monument encircled by a wrought-iron fence. On top of the monument are three bronze figures. General Anthony Wayne is in the center. On his left is an unnamed frontiersman, and on his right an unnamed Native American. General Wayne and the frontiersman are looking slightly to their left; the Native American is looking to his right. All three of them are generally facing the edge of the cliff. The area over which the bronzed warriors gaze is a low prairie of wildflowers and tallgrass that stretches from the base of the bluff a couple hundred yards to the river. North River Road cuts across the plains just before the water. Between the two lives a wetland.

I took a paved trail around the edge of the bluff, down to the lowlands, and wandered around paths that had been mown into the prairie. The grass was taller than my head. The day was hot, and there was no shade. I walked twenty minutes in one direction and found only more grass and flowers. More path. So I turned around. Had I been working toward a clear destination, this would have been called getting lost; as it was, let's call it research.

It was near this place that General Wayne's army met with a volley of musket fire. An advance guard of US scouts had been ambushed by a band of Native soldiers. Six scouts were killed. The rest beat a path back upriver, and ran blunderously into their comrades. Recovering, a troop of dragoons, armed with sabers, charged onto the prairie, followed closely by the bulk of the infantry division. Wayne was traveling in a middle column some ways back. Upon hearing of the ambush, he said, "Prepare to receive the Enemy in front in two Lines." His men obeyed, and suffered the Miami, the Shawnee, the Wyandotte, and some of their Canadian allies above the bluff—a quarter mile or less from where Wayne's likeness stands years later.

What must this have sounded like? Gunshots; shouting; bugles; drums. Howitzers bellowing: *Waugh! Waugh!* Dying men screaming, horses braying. The low, guttural battle cry of the Wyandottes, which some confused for the sound of bells. I stood still among the meadow grass, listened intently, heard only the humming of a breeze.

At Fallen Timbers, the confederacy was outnumbered in the extreme, maybe as much as six to one. Upon the arrival of Wayne's soldiers, they were strung out in a long front that stretched northwest from the banks of the Maumee River, disappearing at the other end into thick woods. The confederacy was fond of this formation. The ends of the line could wrap around enemy armies, obliging the opposition to fight on their sides and rear, as well as in front. A drawback at Fallen Timbers was that, given the Native forces' meager numbers, the line was too thin to withstand any concentrated attack. The Legion of the United States rallied and charged. Wayne sent a Captain MisCampbell in a flanking maneuver to the left, along the bluff. Captain MisCampbell rode straight into a group of Native warriors and was shot in the chest. His men turned back. Most of the force was applied toward the opposition along the river. The confederacy was forced into retreat. They fell back as far as Fort Miami, a British outpost some four miles downstream, where they clamored for sanctuary. The British, in a move that would embitter tribal leaders for decades to come, refused to open the gates.

Wayne wrote to Henry Knox, who was secretary of war at the time, that "this horde of savages, with their allies, abandoned themselves to flight, and dispersed with terror and dismay, leaving our victorious army in full and quiet possession of the battlefield." General Wilkinson disagreed with Wayne's declaration of victory. He then wrote, flatly, "this affair does not deserve the name of a Battle." Some historians call it a skirmish. The whole event had taken less than an hour and a half. A couple of years earlier, a tornado had knocked down hundreds of trees near the prairie on which the fight began. The fallen timbers lent important cover to retreating Native warriors. Eventually, they also lent their name.

The United States did not produce a standard-issue military gun until 1795, so in 1794 Wayne's army was using French muskets from the American Revolution, and US copycats of those muskets. Native tribes, having been armed by the French and British, possessed similar weapons, with the exception that the French would not provide cannons in case allegiances were to switch sides. But muskets they were good for. General Wayne himself was partial to the bayonet. He knew that the success of Native armies was due in part to their strategic uses of close combat and the element of surprise, and believed that it was in hand-to-hand combat that battles between the US and the confederacy would be won. Bayonets were iron blades mounted on the end of a soldier's musket. They were more useful than the muskets themselves in close quarters. Later, Wayne praised the bayonet in a letter: "The bayonet is the most proper instrument, for removing the Film from the Eyes—& for opening the Ears of the Savages, that has ever been discover'd—it has also an other powerful quality! Its glitter instantly dispeled the darkness, & let in the light."

Some Native leaders refused to see "the light" Wayne so jubilantly described. Tecumseh was one. Tecumseh was born in 1771, when this country belonged to Native tribes—and vice versa—and lived the rest of his life as if this were always true. He was Shawnee, and his name in English means "Shooting Star." At the Battle of Fallen Timbers, Tecumseh fought vigorously, wildly, running up and down the battlefield. Right in the middle of everything that would happen that day, Tecumseh accidentally misloaded his rifle, and had to fight for a while with a small-bore fowling piece. Tecumseh was a younger man than Little Turtle or Blue Jacket, and less inclined to accept defeat. When it came time to sign the Treaty of Greenville, Tecumseh refused. At Fallen Timbers, Tecumseh had been among the first to act, crouched waiting in the grass near the edge of the bluff. It was there, overlooking the banks of the Maumee, "where tall meadow grass was interspersed with thin timber," that Tecumseh and his warriors had stood to fire, and where the warriors of the United States had eventually outflanked the Native sequence and forced retreat.

Emerging from the tallgrass prairie, I walked on in the direction of Fort Miami, toward Toledo. I took a trail that passed through Side Cut Metropark, so named because it follows loosely the side cut of the Miami and Erie Canal that once connected it to the city of Maumee. Three of the original locks still stand. During the Depression, laborers with the Works Progress Administration built a couple of stone shelters in the park. I followed a trail through marsh and forest. It was unbelievably green. There was a softness to the place which seemed to absorb every sound. Songbirds flitted across the path. I stopped short—three deer stood grazing not five yards away. For seven minutes they continued, chewing and roving as if I were not there. For seven minutes I stood perfectly still.

In the city of Maumee-Perrysburg, I walked southeast to the outskirts of town, across the river, through a long farmer's market where I bought and ate a dripping peach, and toward the home of Lisa and Chris Roseman. Lisa is my wife's cousin. Arriving, I took a hot shower to rinse off the river, and then we all went out for pizza and to catch up. We talked mostly about my trip along the Maumee, which Chris referred to as "the Mighty Maumee." Chris is a professor at the University of Toledo. After dinner, we drove to the campus so they could show me the Ottawa River, which runs right through the middle of it. This was the first time in seven days I had been outside of the Maumee's watershed. The Ottawa River drains its own small watershed in northwest Ohio. It does not feed into the Maumee, but empties independently into Lake Erie. In 1991, the Toledo Health Department posted signs on campus that read, "Due to water pollution, this area of the river is unsafe for swimming, skiing, other water activities and fishing."

"The university just finished a big restoration project on the Ottawa," Chris said. "It involved a lot of students and faculty. It's been about a ten-year process. There is new signage and a riverwalk through campus. The river has been low this year, but it's a lot cleaner than it was. It's nice to walk around campus in the fall. I plan on taking this way to meetings. They're putting a lot of effort into beautifying the campus, and adding things like this bridge we're standing on."

We drove back to Maumee-Perrysburg, across the Maumee River on I-75. I could see from the bridge Anderson's Grain and Ethanol, one of three major shipping ports in the city, and a Pilkington (glass) manufacturing facility. I closed my eyes. The drone of the car over the bridge was putting me to sleep. When we arrived back at Chris and Lisa's house, I went almost immediately to bed. The next morning, Chris dropped me off next to the river and I walked into Toledo.

———————

Toledo, Ohio, surrounds the mouth of the Maumee on the shores of Lake Erie. Lake Erie is what remains of the Erie Lobe of the Wisconsonian ice sheet. It is the second-smallest of the Great Lakes, after Lake Ontario, and the shallowest—210 feet at its deepest point. Its average depth is sixty-two feet. In the western basin, near Toledo, the lake averages just thirty feet in depth. Because it is shallow, and because it is the southernmost of the Great Lakes, it is also the most hospitable to life, historically speaking. Shallow waters are generally warmer than deep because sunlight heats further down. Also, Lake Erie is full of nutrients from the soils that wash in with rain and melting snow. Most of the nutrients on the west end of Lake Erie come in at Toledo, through the mouth of the Maumee River. These include nitrogen and phosphorous in large amounts—3,800 metric tons per year of phosphorous alone. The formal term for this is "nutrient loading," and it has a lot to do with the algae blooms forming in Lake Erie's western basin.

Algae are basically aquatic plants that thrive on nutrients—mainly nitrogen and phosphorous—and sunlight. Some ecologists call algae "grasses of the waters." Algae can be green, brown, blue-green, red, or yellow-brown; you can see their color when they clump together. They have names like *Spirogyra, Oedogonium, Stiegeoclonium, Ulothrix, Cladophora, Zygnema,* and *Prasiola,* also known as stream lettuce. In a healthy stream ecosystem, algae form the base of the food web. They float freely, invisible to the eye, or gather on rocks and logs, photosynthesizing as plants do. You have probably seen green or brown algae if you've ever picked a rock out of a creek. Most algae, especially forms of green algae, contribute oxygen to a stream, and when they die, green algae decompose into what is sometimes referred to as FPOM: Fine Particular Organic Matter. FPOM is a major source of food for other water organisms. Algae are generally considered to be good citizens in the riverine community.

Up along River Road, past Fort Miami, the Toledo Country Club, more long, shallow lawns, the Maumee River Yacht Club, River Side Marina (next door to the Toledo Sailing Club), and the aquarium of the Toledo Zoo, I had a lot of time to think. Near the Maumee River Yacht Club, I thought about yachts. Near the Toledo Zoo aquarium, I thought about aquariums. It's a short train of thought from aquariums to algae, so after the aquarium I thought about algae.

What is blooming across the surface of Lake Erie is not, strictly speaking, algae. It is cyanobacteria. Cyanobacterium is one of the oldest living organisms on the planet. It used to be classified with algae, colored blue-green, but not anymore. It was eventually reclassified with bacteria, colored cyan. But cyanobacteria are still often lumped in with algae, because they are photosynthetic and aquatic. Air, warm temperatures, and an abundance of nutrients, especially phosphorous, are the perfect conditions for both algae and cyanobacteria to thrive. This is, incidentally, a good description of the western basin of Lake Erie.

Masses of cyanobacteria began to form in Lake Erie around the mid-1990s. They stayed mostly on the surface, near air and sunlight. For twenty years they have persisted, growing progressively worse. Cyanobacteria can draw nitrogen out of the air, which gives it a competitive edge over brown and green algae. Beginning in the 1990s, a thick mat of cyanobacteria grew to shade everything below it. Suddenly, Lake Erie was not so warm and hospitable. Suddenly, lumps of dead blue-green algae sank to the bottom and consumed enormous amounts of oxygen while they decomposed. There was almost no oxygen left for other forms of life. The technical term for this condition is eutrophication. When oxygen levels get as low as they have in Lake Erie, the term is hypereutrophication.

There are some important differences between algae and cyanobacteria. For example, that cyanobacteria can produce a family of toxins called microsystins. Most microsystins are hepatotoxins, which poison animals in the liver. Some species of cyanobacteria also produce neurotoxins, cytotoxins, and endotoxins. Several cases have been documented of human poisoning, usually after drinking or swimming in waters that hosted a bloom. In 2015, the EPA warned that exposure to cyanobacteria "can result in a wide range of symptoms in humans including skin and eye irritation, fever, headaches, muscle and joint pain, blisters, stomach cramps, diarrhea, vomiting, mouth ulcers, and allergic reactions." In extreme cases, "seizures, liver failure, respiratory arrest, and (rarely) death may occur." Recently, some evidence has suggested that microsystins could function as liver carcinogens in humans.

In 2011, the blooms in Lake Erie spread farther than they ever had before, past Cleveland and into the lake's central basin. In 2013, cyanobacteria overwhelmed a water treatment plant in Ottawa County, just southeast of Toledo, forcing it to be shut down—the first time this had ever happened. In 2014, a state of emergency was declared in the city of Toledo: for three days, citizens were unable to drink anything that came out of a faucet. The *Toledo Blade* ran a photograph of a glass of water dipped from the city's intake crib, thick with cyanobacteria. In a widely publicized relief effort, the Ohio National Guard was called in to deliver cases of bottled water. The 2015 bloom was the largest on record, topping even 2011's in size and producing a thick scum "about the size of New York City," wrote the *Detroit Free Press*, though it stayed near the center of the lake, sparing the water supplies of coastal towns. Wildlife and vegetation were not so fortunate. *National Geographic* declared this to be the "new normal," exacerbated by climate change and irresponsible land management.

What is happening in Lake Erie is a disaster by nearly any measure—ecologically, economically, socially, culturally. I have watched media coverage of these blooms each year with a growing sense of unease. It is unsettling, as a creature, to watch as other creatures are increasingly unable to live in the places they call home, to notice the world slowly becoming uninhabitable.

———————

Where blooms are concerned, this is round two for Lake Erie. Round one came in the middle of the twentieth century. At that point, people had been dumping sewage and wastewater into the lake and the Maumee River for decades. In the 1950s it caught up with them. The story was told perhaps most eloquently by the mayfly. Mayflies had always swarmed Lake Erie in hordes. Female mayflies would lay their eggs four thousand at a time on the bottom of the lake. Mayfly nymphs lived on lake-bottom oxygen. Every June for thousands of years, adult mayflies would return to the shores of Lake Erie for egg-laying and then a massive die-off. There were so many of them in the 1940s that people would use snowplows to clear away the bodies. For the next dozen or so years, as oxygen in Lake Erie dwindled, mayfly populations did too. In 1956, no mayflies showed up at all.

In the 1960s, people started to pay serious attention. *MacLean's* ran the headline, "Lake Erie is Dying." *Newsweek* called it "the Dead Sea." Reports spread wide of the lake's impending demise. Diagnosis: eutrophication.

As early as the 1890s, biologists and fishermen had noticed the effects of sewage and industrial byproducts, like sawdust, on streams. But there wasn't enough evidence—or knowledge—to make a case that would capture the interest of the general population. By the 1930s, the presence of waste products in rivers was measurable, but early reports deemed it insignificant. It would take another thirty years for pollution to be taken seriously. Meanwhile, in the 1940s, Thomas Huxley Langlois, a fishery biologist at Ohio State University, began advancing the idea that the death of organic matter in the western basin of Lake Erie could "withdraw oxygen from the water." One example of this, he mentioned, was blue-green algae. In 1951 and 1952, Kenneth G. Wood, a doctoral student under Langlois, studied invertebrate populations in the basin. Wood concluded that a decline in the lake's environmental conditions would be signaled by an increase in the invertebrate death rate. In August, 1953, zoologist Ralph Dexter of Kent State University took some students out for a routine dredging in western Lake Erie, looking for specimens to examine at Stone Laboratory. When they pulled up their catch, all of the specimens were dead.

Still, outside of a small contingent of scientists and fishermen, most people would not have thought about the effects of waste and sewage on the health of Lake Erie and its non-human residents; such questions were relatively uncommon before the environmental movements of the 1970s. To the extent that they did, the general response was that such an enormous lake—nearly ten thousand miles square—could handle it. That is, until it became undeniably apparent that, in fact, the lake couldn't. Garbage, rotting fish, and decomposing algae kept washing up onto Lake Erie's beaches. You could smell it miles inland. Lake herring, lake sturgeon, whitefish, and northern pike were dying off. Lake trout disappeared. Boats, motoring across the lake, carved through rafts of algae two feet thick and as far as the eye could see. These blooms lived on nutrients from waste—agricultural, industrial, municipal—spilling from the mouth of the Maumee River, which had begun to bear an unhealthy resemblance to a sewer.

In 1965, the Junior League of Toledo produced a documentary film in collaboration with WGTE, the local PBS affiliate television station. The documentary was called *Fate of a River: Apathy or Action*. It was a twenty-eight-minute whirlwind tour of deteriorating conditions in the Maumee River and its watershed. The film featured grainy footage of dirty beaches, gasping fishes, and algae-laden streams. A narrator outlined the influence on the Maumee and its tributaries of chemical plants; steel fabrication; untreated wastewater; heated water discharge; raw human sewage; and streamside dumping in floodplains. The narrator said, "Nature strives to cover the earth; man uncovers it and sends silt down the streams." The film described waterways covered in "great masses of oil"; "the scum of stagnation"; "violent discolorations"; "white detergents"; and algae that "looks as if it would take a plow to create a channel in it." It described the Maumee River itself as "inky black," and "burdened with the debris of a dozen cities." In the film's final line, as the music crescendoed, the narrator concluded: "The fate of our rivers is in our hands. All hinges upon our apathy or action."

Seventy thousand people saw the film. Volunteers held screenings at libraries, garden clubs, Rotary clubs, bridge clubs, in schools, and at community board meetings. It was only three years since the publication of Rachel Carson's *Silent Spring*, and there was a lot of energy around issues of the environment. The women of the Junior League captured this energy and pointed it toward the Maumee. A scenic byway was created. Cities and counties built wastewater treatment plants along the river. Fishing was restored in some areas previously judged too poisonous, and bald eagles returned after a thirty-year absence. Some members of the Junior League of Toledo were invited to testify before a Congressional committee debating the Clean Water Act of 1972.

Nutrient loading, including that from the Maumee River, was one of the main reasons anybody was even considering a Clean Water Act. In the region's manufacturing heyday, runoff and wastewater laced with chemicals and phosphorous poured into the river. One major source was the City of Toledo. A 1975 report commissioned by the EPA cited "a large cooling-water discharge from the Acme powerplant," as well as "the erratic performance of Toledo's sewage treatment plant" among the reasons for the river's "degraded" status. The report summary concluded: "Even if Toledo were to be wiped off the map, these conditions would not entirely disappear, nor would many of them be much changed."

―――――――――――――

Toledo is still there. Today, it is a poster city for the Rust Belt, a term which, loosely defined, encompasses any upper midwestern city built on industrial manufacturing during the late nineteenth and early twentieth centuries (and subsequently affected by a decline in manufacturing across the region), referring to the tendency of uncared-for metal to rust. The process by which rust forms is called oxidation. It is a chemical reaction between ferrous metal—steel, stainless steel, wrought iron, cast iron—and water. More specifically, it's a corrosive reaction between iron and oxygen, facilitated by water. In places that do not have much water, like deserts, there is no rust; without lakes and rivers there would be no Rust Belt.

Toledo has, perhaps more than anything else, an abundance of water and metal. Taking Broadway Street through the city's south side neighborhoods, I shortcut the last major bend in the Maumee and entered what appeared to be an old manufacturing district. At the Amtrak rail yards, I hung a left, then right, and found myself among a row of nondescript buildings, brick and cinder block. Empty lots, all gravel and weeds, were protected by chain link and barbed wire. Windows were boarded up or bricked over. Occasionally a vague sign: something like, "JJ Supply Co." JJ Supply Co. is a wholesale distributor for industrial pipe, valves (ball valves, gate valves, check valves, butterfly valves, plug valves), flanges, and fittings. It is still in operation. Toledo businesses that are no longer in operation include smelting factories, responsible for extracting iron from ore. This was done by heating the ore past its melting point, and putting it into contact with an oxidizing agent, like air. Smelting fires in Toledo were fueled by charcoal shipped down the Maumee from the Black Swamp.

In the twentieth century, Toledo became more famous for its glass manufacturing industry than for metal, earning the nickname "Glass City." But glass is a product, while metal, in modern industrial cities, is infrastructure. There continue to be steel fabricators, metal handlers, and suppliers in Toledo, with names like Art Iron and Alro Steel. In the early years of the twenty-first century, ironworkers, along with cement finishers and other laborers, fashioned the components for an enormous bridge—to cross the Maumee River—on Front Street in Toledo, on a site where iron ore had once been smelted.

Shot through, as it is, with the Maumee River, and sitting on a popular port, Toledo has always been involved with shipping and transportation. Early on, it was a leader in sailing ship production, sending the lumber of the Great Black Swamp across the Great Lakes. Now, four major freight railroads intersect at Toledo: Canadian National, CSX, Norfolk-Southern, and Wheeling and Lake Erie. Over the years there have been many more, including the Toledo, Wabash and Western; the Detroit, Monroe, and Toledo; the Toledo, St. Louis, and Kansas City; the Ohio Central; and the Toledo Terminal Railroad, which made a loop around the city, crossing the Maumee twice. Toledo was also an early entrant into automobile production. The city continues to make Jeeps, the Wrangler and the Liberty. It also still has shipyards near the Port of

Toledo. The Port of Toledo is a major port. Its biggest commodities are coal and iron ore.

In 1868, a Toledoan named Jesup W. Scott published a pamphlet titled, *Toledo: Future Great City of the World*. His first name was misspelled as Jessup, with two s's, on the cover. Scott argued in this pamphlet that geography and context are prime shapers of a city's success. Water was at the center of his vision. "The cities of Lake Erie will show, and permanently maintain, the more rapid ratio of growth," he wrote. "One of them will, in the end rival and dominate Chicago." In Scott's mind, this would obviously be Toledo. Four years after the publication of *Toledo: Future Great City of the World*, he founded the University of Toledo in the interest of equipping the city for its inevitable future greatness. Near the end of his 1868 treatise, Scott wrote, "One hundred years to come, with the command of steam, electricity, and we know not what other and superior agencies for wonder-working, can scarcely fail to produce results of a magnitude beyond the power of the most vigorous imagination to conceive."

One hundred years later, Toledo was an industrial heavyweight surrounding a polluted river on the edge of a terminal lake. In the late sixties and into the seventies, algae blooms were still common in Lake Erie. In 1972, the United States and Canada signed the Great Lakes Water Quality Agreement. This cut the amount of phosphate entering Lake Erie from both Canada and the United States from twenty thousand tons per year to eleven thousand. The improvement was dramatic. By 1987, when the Great Lakes Water Quality Agreement was revisited, the nutrient load in Lake Erie was down to thirteen thousand tons. The mayfly had returned. Algae blooms had all but disappeared—for a few years, anyway. By the middle 1990s they were back.

The blooms of the 1990s differed from those of the sixties and seventies in that they formed closer to the shore and contained more cyanobacteria. For a long time, people thought it was mainly phosphorous that spurred the growth of the blooms. This was based to some extent in the fact that cyanobacteria can pull nitrogen from air, and so cutting off the supply from rivers presumably wouldn't have any effect. More recently, though, scientists like Drs. Robert W. Howarth of Cornell University, and Roxanne Marino of the Ecosystems Center in Woods Hole, Massachusetts, have argued that both phosphorous and nitrogen will have to be decreased. Both nutrients came—and continue to come—primarily from fertilizer use in agriculture, on golf courses, and on residential lawns. In urban areas sewage is the culprit.

In the case of Toledo, there are no city records documenting the construction of the earliest sewers, but it is believed that they were brick, and simply poured into the Maumee without treatment. Sewage treatment began in the 1930s, with settling tanks and chlorination facilities. Flocculation began in the 1950s. The whole system was redesigned to combine sewage and groundwater runoff into a single sewer. This worked well for the most part, except that if it rained hard enough the system would overflow into the Maumee River. In 2002, the citizens of Toledo approved the Toledo Waterways Initiative, a system of forty-five separate projects which are collectively intended to cut 470 million gallons of overflow per year by 2020.

Back at the headwaters, Fort Wayne has been having its own septic issues. A 2008 federal mandate required the city to decrease the amount of raw sewage it was sending downstream. By the time I moved to the banks of the Maumee in the fall of 2013, untreated waste entered all three rivers, and several smaller tributaries, at forty-three different points in the city. The current goal is to reduce the city's collective dump from 1 billion gallons to 100 million gallons annually by 2025. In February of 2016, the city of Fort Wayne announced plans for a project called the Deep Rock tunnel, a five-mile-long, sixteen-foot-wide cement pipe, to be buried two hundred feet deep and under the city. Its purpose is to relieve pressure from the combined-overflow sewer system. Construction of the Deep Rock tunnel will take three years. It will be cut by a tunnel boring machine, or TBM. Construction of other related piping and infrastructure will take an additional five years. At least one pipe will cross underneath the Maumee River. The Deep Rock tunnel will be the biggest public works project in Fort Wayne history; once in operation, it is expected to decrease sewage overflow by 90 percent. The whole system is scheduled to be completed sometime in 2025.

———————————

The day had grown overcast and was indicating rain. I pulled my jacket out of my pack, put it on, and went looking for someplace indoors to eat a late lunch. Crossing Swan Creek, a feeder stream of the Maumee that enters the river four and a half miles from the end, I reached the south side of downtown and found a small café between a flower shop and a sushi bar. At the counter I ordered a sandwich and a green tea. There were two baristas working. I asked them if they had heard anything about an algae bloom showing up in Lake Erie.

One of the baristas, a tall, slim woman in a plaid shirt, said, "It's not here yet, but we're expecting it any time."

Here we were—four thousand years, at least, after the formation of the Maumee River; more than two hundred years after white settlers violently wrested control of the basin; one hundred and fifty years after the draining of the Great Black Swamp; one hundred and forty-six years after Jesup W. Scott called Toledo a "Future Great City of the World"; seventy years after algae blooms killed off the mayflies in Lake Erie; fifty-one years after the Junior League of Toledo warned of the Maumee River's declining health; twenty-six years after cyanobacteria re-emerged in Lake Erie's western basin; and eight years after a federal mandate required the city of Fort Wayne to decrease its toxic dump more than 90 percent—standing along the edge of the Maumee River, waiting on catastrophe.

CHAPTER 7

Sometime in the middle of the afternoon, I realized I was going to reach the mouth of the Maumee River that day. The plan had been to walk a dozen miles into Toledo, find a hotel, and make the last leg of the trip—four miles or so—the next morning. But since I had about five hours of daylight left, and since the only thing on my agenda for that day was "walk through Toledo and look around," I pulled out my map and decided a route to the mouth.

Once, Ohio and Michigan fought a war over the mouth of the Maumee River. It started with a boundary line run and marked by surveyor William Harris in 1817. There had been some uncertainty about the northern boundary of Ohio, and the Harris Line, as it came to be called, was commissioned by Congress to settle the matter. The Harris Line was marked due west from Lake Erie, beginning at a point on the north bank of Maumee Bay. It was a fine line in nearly every respect, except that it cut through land that, according to the Land Ordinance of 1787, belonged to the territory of Michigan, the Toledo Strip. The Toledo Strip was five miles wide at one end, eight on the other. It ran along the southern edge of the Ohio-Michigan border, west from—and including—Maumee Bay. In 1818, an alternative line was run by John Fulton, at the urging of Michigan Governor Lewis Cass. The Fulton Line intersected Lake Erie southeast of Maumee Bay, locating it in Michigan Territory.

For a couple of decades, the Toledo Strip was in limbo. Then, in 1835, Ohio went ahead and officially declared the Harris Line to be the northern boundary of its state—and decided in the process to re-mark it for good measure. In response, the legislature of the Michigan Territory passed "An act to prevent the organization of a foreign jurisdiction within the limits of the Territory of Michigan." Anybody caught organizing a foreign jurisdiction—like, for example, extending the boundaries of Ohio—could be fined a thousand dollars and sentenced to five years' hard labor. When Ohioans showed up to re-draw the Harris Line in 1835, nine of them were arrested.

There's nothing like politics to make people choose sides. The Michigan Territory, emboldened, spent the spring and summer of 1835 arresting Ohio partisans in Toledo. Probably one of the most interesting arrestees was Benjamin Franklin Stickney. Benjamin Franklin Stickney was a former Indian agent in Fort Wayne who had bought land near the mouth of the Maumee. Stickney was an opportunist. He was initially in favor of Michigan authority so he wouldn't have to pay Ohio state taxes, but switched sides when he realized that it would profit him more to have Ohio build a canal through the area. (Canal access to Lake Erie was among the main reasons Ohio wanted Maumee Bay.) Stickney was also an eccentric who, in the words of historian Tana Mosier Porter, "often experienced problems in getting along with people." He named his sons One and Two, in order of birth, believing that male children ought to have the opportunity to choose what they would be called. He named a daughter Indiana. His sons never changed their names. Benjamin Franklin Stickney was arrested by Michigan law enforcement on at least three occasions in the summer of 1935. On July 14, 1935, Two Stickney was put under arrest in Toledo by Joseph Wood, deputy sheriff of Michigan's Monroe County. Two Stickney resisted, spearing Wood with a knife blade between the second and third ribs, and shouting, "There, damn you, you have got it now!" Wood at first believed this to be a mortal wound, but he survived. Two Stickney escaped.

Three days previous, Governor Robert Lucas of Ohio had directed an Ohio judge to hold court in the Toledo Strip. The official term was "exercise jurisdiction." In the eyes of the federal government, exercising jurisdiction would call the strip Ohio. The court session was supposed to be on September 7, 1835. Hearing of the plan, Governor Stephen T. Mason of the Michigan Territory set out to disrupt the event. He gathered together a group of armed men to ride into Toledo and prevent the court from doing anything. When they arrived on the seventh, nobody was around. Ohio had outfoxed Mason by sneaking into Toledo at one o'clock in the morning and holding court, by candlelight, around three. After the ceremony, the Ohioans stopped for a celebratory drink, then beat it south along the Maumee River toward established Ohio territory. The clerk's notes, the official record of the session, were tucked into the hat of the clerk, Dr. Horatio Conant. At some point along the way Dr. Conant got his hat knocked off by a tree branch, and the party had to scour the woods for the notes. Eventually they were found. Conant and crew hurried them safely the rest of the way to the town of Maumee. And all before breakfast.

Relations between the governments of Ohio and Michigan, which were already precarious, became more so. In June of 1836, President Jackson, trying to avoid a war between states, signed the Northern Ohio Boundary bill, which offered Michigan a deal: give up the Toledo Strip and in return get statehood, plus an additional section of land to the north, across the Straits of Mackinac. That winter, Michigan agreed to Jackson's terms. The Toledo War was over.

Yet it feels somehow exaggerated to call what happened between Ohio and Michigan a war. Though warning shots were fired on at least one occasion, and though at least one person was nearly tarred and feathered, and though some individuals—including Andrew Palmer, Ohio partisan and editor of the *Toledo Gazette*—were regularly harassed, and though Two Stickney had stabbed Joseph Wood in between the second and third ribs, causing him to bleed excessively, and though the governor of Michigan did assemble a militia and march on Toledo, the conflict had no fatalities. (With one exception: a horse was killed when Michiganders fired on a barn which they thought housed a store of weapons.) It was a political battle more than military, fought with courts and acts and scathing-yet-politely-worded letters. The essence of the argument was control of the harbor on the Maumee, and of the farmland north of the river. There was also a good deal of spite. Ultimately, Ohio got the Toledo Strip and Michigan got the Upper Peninsula.

Leaving the café, I walked up Water Street, along a cement riverwalk, under the Veterans' Glass City Skyway, and into a stretch of riverfront plots with miscellaneous commercial uses—grain elevators, power substations, a garbage truck corral, and a slug-like heap of salt four hundred feet long, covered with black tarpaulins. I stopped to rest a few minutes at Jamie Farr Park, then continued on. More grain bins, manufacturing plants, power towers. The sidewalk was overgrown. Across a set of railroad tracks, past barbed-wire fence and plywood windows, I curved left on Summit Street, around the rambling Division of Water Reclamation. I was nearly there. My route cut through Bay View Retirees Golf Course. A white ball bounced across the path in front of me, and two older men in a cart bounced after the ball. Beyond the golf course and around the Bay View Toledo Yacht Club, I came, finally, to Bay View Park.

Bay View Park may be the conclusion of the Maumee River; it's hard to tell. The end of the Maumee is as elusive as the beginning. This park was the closest thing I could find. Really, it's a forest. A trail blazed grey through the woods, and I followed. A sign: "BIRD

WATCHING TRAIL." I dutifully watched as birds hopped across the gravel, darted through the air. Trunks pillared along the edges of the path. Undergrowth filled in the gaps, and ragweed shone the way in gold. Around a loop, the trees gave way and there it was: the mouth of the Maumee River. I had arrived—110 miles later. Standing at the water's edge, I looked out past a frame of brush. Water poured in over itself—river becoming bay. A wooly island in the middle. On the opposite shore: infrastructure, brown and metal. To my left a long break. Beyond, Lake Erie.

In 1922, Ohio extended the Harris Line east, across the lake, until it hit an international border somewhere north of Sandusky. Michigan said, "Okay." Ten years later, Ohio decided to re-draw the line at a forty-five-degree angle to the northeast. The new boundary gained them an extra two hundred square miles of lake, called the Erie Triangle. Michigan was fine with this too, until 1947, when oil and gas were found in the area. Then, Michigan claimed ownership of the Erie Triangle. In 1952, the United States Geological Survey went with the forty-five-degree line. In 1965, as an expression of goodwill, the governors of Ohio and Michigan met for a photo op near the easternmost landed point of the Harris Line. One year later, Michigan filed suit against Ohio for the Erie Triangle. In 1967, the Supreme Court sided with Ohio. In 1971, Michigan sued again, and a federal judge upheld the ruling of 1967. After a couple of appeals, the Supreme Court finally settled the matter, proclaiming the Erie Triangle to be part of Ohio in 1973.

Sitting at the edge of Maumee Bay—on a black metal bench set back from the shore—I was in position to watch shipping vessels arriving and departing the Port of Toledo. The Toledo Harbor Shipping Channel runs from Lake Erie through Maumee Bay and up into the Maumee River. Sediment pumped in by the river starts to shallow the channel over time. Every year, typically in late August, the Army Corps of Engineers takes a couple of barges and scoops around a million cubic yards of sediment from the bottom. This accounts for 40 percent of the sediment dredged out of the Great Lakes every year. Then they drag the barges out into Lake Erie and dump it all over the side. This kind of open-lake disposal is frowned upon generally. It used to be mainly environmental groups who took issue. Since the shutdown of Toledo's water intake crib, other people have also been thinking about all the fertilizers and pesticides sitting out there in the water. In April 2015, Ohio Governor John Kasich signed a bill banning open-lake dumping. By 2020, dredgers will have to find something else to do with the sediment.

So far, they have decided to put some of it back on the ground. To that end, around the time I was walking through Toledo, the Toledo-Lucas County Port Authority opened the Great Lakes Dredged Material Center for Innovation. The Great Lakes Dredged Material Center for Innovation is on the north side of the Maumee River, next to Jamie Farr Park. It consists of four 2.5-acre "cells" of land, where, between August and September, fifty thousand cubic yards of sediment were put. It used to consist of Riverside Park. Over the next two years, compost will be added to the silt, and corn and soybeans will be planted in the mixture. If the experiment works out, more sediment will be used to level farm fields across the basin. The idea is that all this dirt being pulled out of the water used to be farmland, and it might as well be that again. Craig Butler, the director of the Ohio EPA, said in August 2016, "The best soil in Ohio sits at the bottom of Lake Erie."

––––––––––

Currently, nobody is fighting a war in the watershed of the Maumee River. But neither is there peace. Outside of the terror of cyanobacteria, the relationship between people and the river is strained. Floods drown cropland and bike paths. Boaters and anglers use the river, but at arm's length. Soil and pollutants continue to pour in from yards, fields, and factories, and stormwater overflows after hard rains. The dead zones in Lake Erie are the most recent example of the kind of influence people have had on this region, but they are by no means the extent of it. Probably the truest thing you could say about the Maumee River is that it exists entirely in relationship with everything around it. Probably the second truest thing you could say is that for the last two hundred or so years we have mainly operated as if this were not the case.

I do not believe that there is no harmony to be found between living things, despite so much evidence to the contrary. It is a stubbornness—a hope—I was taught by the river. A growing body of research suggests that a shift to more sustainable farming and lawn maintenance practices, coupled with a reintroduction of wetlands along the river, can help cut down on pollution and erosion. Experimental wetland restoration projects elsewhere in the Midwest, and some along the Maumee, have been successful in reducing the amount of topsoil, phosphorous, and nitrogen flowing into rivers. Ray Stewart of the Ohio Wetlands Association wrote in 2015 that "the return of natural functions in the basin will dampen the negative impacts that centuries of alteration have exacted," and are an important step toward reversing consequences like Lake Erie's algae blooms. It turns out that, given a chance, waterways can regain their balance. It is possible that we could learn at last how to live in a watershed.

I stayed for a long while staring out at the mouth of the Maumee River. Across the water the tall arm of a crane, black with orange on top, reached up and down. The difference between the green park where I sat and the south side of the river—a clutter of docks and machinery, a smokestack projecting into the sky— was conspicuous. Two visions of the future. Between them the Maumee, stretching back through time. A bird, a yellow-bellied flycatcher, hopped over near the bench. My notebook sat open on the seat. I realized that I had been expecting something to happen here. Enlightenment, maybe. Whatever it was never came; or not like I thought it would. As the sun burned low in the western sky, and the day took on the gold cast of evening, I stood, turned around, and walked back upstream. There is no one great lesson here, only a thousand small ones: the rocks, the trees, the cities, the fields—and this holy water, coursing through us all.

NOTES

CHAPTER 1

pp. 11-12: Geographical and geological information on the Maumee River and its watershed comes from maps and documents produced by the United States Geological Service and the Ohio Department of Natural Resources, especially the Office of Coastal Management. Details on the length of the river and size of the watershed are on the DNR website at coastal. ohiodnr.gov/maumeeriver. Other sources put the size of the basin even larger, up to 8,300 square miles.

p. 12: I read about the algae blooms in Lake Erie in dozens of periodical articles over the past few years, too many to name. The most consistent coverage has come from the *Toledo Blade*. Other regular sources are the *Detroit Free Press* and *National Geographic.*

p. 12: Efforts are underway to make the rivers more visible to Fort Wayne residents. A large-scale riverfront development project is planned on the banks of the St. Marys River. As of this book's completion, properties had been acquired and designs had been approved for the first phase of the project. (See the Fort Wayne *Journal Gazette,* "Riverfront Fort Wayne gets Funding Approval" [Dave Gong, April 11, 2017], and "Board of Park Commissions Closes on Final Three Properties for Riverfront Development" on the WBOI website [March 10, 2017].)

p. 13: Floods are mentioned in Fort Wayne newspapers—the *Journal Gazette* and the *News-Sentinel*—and other places.

Two of the most useful articles were "6 Deadly Days: How the Flood of 1913 Devastated Fort Wayne," by Cindy Larson in the *News-Sentinel* (March 20, 2013), and "In Fort Wayne, a Strain on Flood Dikes and People," in the *New York Times* (March 18, 1982).

p. 13: Robert Sayre's quote comes from page 101 of "Learning the Iowa River" in the *Iowa Review* (Fall 2009).

p. 14: The definition of 'stream' is in the US Board on Geographic Names's Geographic Names Information System (GNIS).

p. 15: My discussion of place names in this chapter comes from *Native American Place Names of Indiana* by Michael McCafferty (University of Illinois Press, 2008), pages 81-82 (on Maumee), and pages 79-81 (Kekionga). That people long thought Kekionga meant "blackberry bush" is mentioned in a blog by the Allen County Public Library's Genealogy Center ("The Origins of 'Kekionga' in Fort Wayne's Past, Pt. 1"). The definition of Kekionga as "hair-clipping place" comes from an article in the October 20, 1990 edition of the *News-Sentinel*, titled "Little Turtle's Famed Battle Humbled U.S. Forces." That the Maumee was called the Miami of the Lake or Miami of Lake Erie is on page 7 of *The Facts and Historical Events of the Toledo War of 1835* by Willard Way.

p. 16: The route from Lake Erie to the Mississippi over which traders would travel was called the "Wabash Trade Route." One historian notes that the route was used so heavily that it became "as well worn as a muskrat slide…" (See "History of Canals in Indiana" by Howard Payne Comstock in the *Indiana Quarterly Magazine of History* [Vol. 7, No. 1, March 1911]; the muskrat quote is on page 1 of that article.)

p. 16: Information on Miami life comes from *The Miami Indians of Indiana: A Persistent People, 1654-1954* by Stewart Rafert (Indiana Historical Society, 1996), and *The Miami Indians* by Bert Anson (University of Oklahoma Press, 1970). Anson writes on page 32, "...the Miami were astride the best route between Canada and Louisiana."

Little Turtle's "Glorious Gate" quote is on page 25 of Rafert, and on page 61 of George Cottman's "The Wabash and Its Valley: Part I—The Earlier History" in the *Indiana Magazine of History* (Vol. 1, No. 2, 1905).

p. 16: There are many different spellings of Little Turtle's Miami name. Mishikinakwa is one of the most common, as is Me-She-Kin-No-Quah, which is used on his grave marker, and Meshekunnoghquoh, which is the phonetic interpretation listed next to his signature on the Treaty of Greenville. The spelling used here is the one used by linguist and Miami Tribe of Oklahoma member Daryl Baldwin and others in *myaamiaki aancihsaaciki: A Cultural Exploration of the Myaamia Removal Route*, a document prepared for the Miami Tribe of Oklahoma Cultural Resources Office in 2011 by George Strack, George Ironstrack, Daryl Baldwin, Kristina Fox, Julie Olds, Robbyn Abbitt, and Melissa Rinehart.

pp. 17-18: Information on the St. Joseph River Dam and the Three Rivers Water Filtration Plant comes from a fact sheet on the city of Fort Wayne's website. Information on the development of Fort Wayne's water systems is drawn largely from Tom Castaldi's article, "Mayor Hosey's Three Rivers Water Plant," originally published in *Fort Wayne Magazine* of July 2008.

Details on the filtration plant's capacity are in the 2016 Annual Drinking Water Quality Report from the city of Fort Wayne.

p. 18: The first dam on the Maumee River is the Hosey Dam, named for the city's former mayor. It is a "low-head" dam. Low-head dams are known to produce dangerous hydraulic backwash currents. In June of 2016, a kayaker from Fort Wayne named Sean Hiebel was pulled over the dam and tragically drowned (see "Victim in River Identified as Missing Kayaker" in the *Journal Gazette* of July 7, 2015, and "New Signs Will Promote Safety on Rivers" in the *News-Sentinel*, May 25, 2016).

p. 20: "Another Algal Bloom Found in Maumee River" is the headline of a story by WANE-TV that ran on July 28, 2016.

CHAPTER 2

pp. 21-22: The quote on river length is in "USGS Water Fact Sheet: Largest Rivers in the United States" (Open File Report 87-242, May 1990 by J.C. Kammeyer).

pp. 22-24: Information regarding glaciers in general, and the Laurentide ice sheet in particular, comes mainly from *Great Lakes: Natural History of a Changing Region* by Wayne Grady (Greystone Books, 2007), and from other maps and articles too many to name. The quote from Agassiz, on "God's Great Plough," comes from Great Lakes, page 64.

Louis Agassiz was one of the first Europeans to talk seriously about glaciers. According to biographer Christoph

Irmscher, the concept of mass glaciation excited Agassiz primarily because it seemed to reinforce his worldview of "catastrophism," in which a divine dreator periodically intervened in the world to remake it. (See *Louis Agassiz: Creator of American Science*, 2013.) Agassiz believed that catastrophism refuted the developing science of evolution; if an ice age killed all life on earth, he reasoned, modern species could not be descendants of premodern creatures.

His commitment to distinctive origins extended to humans. In a 2001 article titled "Morton, Agassiz, and the Origins of Scientific Racism in the United States," Louis Menand writes that in 1846, around the time that his ideas about glaciers were gaining traction, Agassiz claimed in a lecture that "although Negroes and whites belonged to the same species, they had separate origins." Agassiz is an outstanding example of the capacity of humans to be both extraordinarily brilliant in one area and profoundly lacking in another.

p. 24: Information on Lake Maumee and the Maumee Torrent came from the Indiana Geological Survey, in a document titled "Surficial Geology," compiled by Anthony Fleming and Robin Rupp; from "The History of Lake Erie," in the Ohio Department of Natural Resources's Ohio Geology Newsletter, Fall 1989; and from an online document by the Little River Wetlands Project, titled "Geology of the Little River Valley." The quote "overtopped a sag" appears in at least two of these publications: from the Indiana Geological Survey and the Little River Wetlands Project.

p. 26: The Miami origin story and their eventual places of residence are found in a chapter by George Ironstrack, titled "nah meehtohseeniwinki: iilinweeyankwhi need isi

meehtohseeniwiyankwi aatotamankwi: To Live Well: Our
Language and Our Lives," in B*eyond Two Worlds: Critical
Conversations on Language and Power in Native North America*
(Edited by James Joseph Buss and C. Joseph Genetin-Pilawa
[State University Press of New York, 2014]). Also from "A
Myaamia Beginning" at myaamiahistory.wordpress.com (a
community blog run by George Ironstrack of the Myaamia
Center at Miami University in Ohio). The quote "from
time immemorial…" comes from page 94 of "Searching for
Our Talk and Finding Ourselves" by Teresa McCarty, Daryl
Baldwin, George Ironstrack and Julie Olds in *Language
Planning and Policy in Native America: History, Theory, Praxis*
(Multilingual Matters, 2013).

p. 27: The translation project I referenced using Jesuit documents
is ongoing at the Myaamia Center at Miami University in
Ohio. It is directed by Daryl Baldwin.

p. 28: The early maps I reference in this chapter are found in
History of the Maumee River Basin, 1905, by Charles Slocum
(pages 75-80), and on page 62 of Cottman, George S. "The
Wabash and Its Valley: Part I—The Earlier History," *Indiana
Magazine of History* (Vol. 1, No. 2, 1905).

p. 29: The lead plate laid by de Bonnecamps and de Blainville is in
the possession of the Virginia Historical Society.

p. 30: Details on Little Turtle are found in Harvey Lewis Carter, *The
Life and Times of Little Turtle: First Sagamore of the Wabash*
(University of Illinois Press, 1987), in "Chief Little Turtle,
Chief of the Miami," compiled by the Allen County Public
Library, in the Ohio Supreme Court Biography of Little
Turtle, and in assorted other books on the Miami already

mentioned. The exchange between Little Turtle and Volney is on page 125 of Love, N.B. "Me-She-Kun-Nogh-Quah, or Little Turtle, 1783-1812," in *Ohio Archaeological and Historical Quarterly* (Vol. XVIII, No. 2, April 1909).

Another example of Little Turtle's wit: In November of 1792, Little Turtle ambushed US. Colonel John Adair, killing six soldiers and capturing 150 horses. Some years later, Little Turtle and Adair met again, and Little Turtle said, jokingly, "A good general is never taken by surprise." (This anecdote is related in a foo note on page 175 of William Heath, "Re-evaluating 'The Fort Wayne Manuscript,'" in the *Indiana Magazine of History*, Vol. 106, No. 2, June 2010.)

pp. 30-33: The overall account given here for the conflict between the US and Native tribes draws primarily from three sources: Alan Brown, "The Role of the Army in Western Settlement: Josiah Harmar's Command, 1785-1790," in *Pennsylvania Magazine of History and Biography* (April 1969); "General Josiah Harmar's Campaign Reconsidered: How the Americans Lost the Battle of Kekionga," in *Indiana Magazine of History*, (March 1987); and *Carter, The Life and Times of Little Turtle.* Harmar's journal entry is mentioned in Warner, on page 49.

p. 33: The Treaty of Greenville is discussed in many of the sources mentioned above. Little Turtle's quote urging his comrades not to fight comes from *Henry Howe, Historical Collections of Ohio*, 1850, page 425, and "Little Turtle's Oration before the Battle of Fallen Timbers" in *Old Fort News*, Vol. 2 (1937). That Little Turtle reminded Wayne, at the treaty of Greenville, of the Miamis' long tenure in the area is on page 61 of Cottman's "The Wabash and Its Valley: Part I" (1905).

pp. 33-34: Settlement information comes from the sources listed
above, and from W.A. Brice, *History of Fort Wayne, From the
Earliest Known Accounts of This Point to the Present Period*
(D.W. Jones & Son, 1868). The ornamental sword given
to Little Turtle by the United States is mentioned in "Chief
Little Turtle, Chief of the Miami."

p. 34: The long quote on Myaamia Removal is found on page 4
of *myaamiaki aancihsaaciki: A Cultural Exploration of the
Myaamia Removal Route* (Miami Tribe of Oklahoma Cultural
Resources Office, 2011).

p. 36: The racial dimensions of Miami removal were underscored
by Lewis Cass at the 1843 Fourth of July Celebration in Fort
Wayne. After accusing the Miami of practicing cannibalism
on their captives, Cass said: "Thank God, the council fire
is extinguished. The impious feast is over, the war dance
is ended; the war song is unsung; the war drum is silent,
and the Indian has departed…today, this last of the race is
here." (This quote is in two separate sources: William Henry
Smith's *The History of the State of Indiana from the Earliest
Explorations by the French to the Present Time*, 1903, page 28;
and Wallace A. Brice's 1868 volume *History of Fort Wayne,
from the Earliest Known Accounts of this Point, to the Present
Period*, pages 123-124.)

CHAPTER 3

p. 39: I learned about meander loops and helicoidal flow from
River Dynamics by Pierre Julien (Cambridge University
Press, 2002); from *Streams: Their Ecology and Life* (2001) by
Colbert Cushing and J. David Allan; and from Lee Sandlin's

discussion in *Wicked River: The Mississippi When It Last Ran Wild* (Vintage Books, 2010).

p. 42: The Black Swamp is described in many sources. Martin R. Kaatz, in "The Black Swamp: A Study in Historical Geography" (*Annals of the Association of American Geography*, Vol. 45, 1955), has a nice roundup. Other useful sources are *Historical Atlas of Paulding County, Ohio* (TME Western Publishing Company, 1892) by Morrow and Bashore, and Nevin Winters' *A History of Northwest Ohio: A Narrative Account of its Historical Progress and Development from the First European Exploration of the Maumee and Sandusky Valleys and the Adjacent Shores of Lake Erie Down to the Present Time* (Lewis Publishing Company, 1917). The trees of the swamp are listed in *Historical Atlas of Paulding County, Ohio* (1892).

pp. 42-43: Early travel difficulties in the Black Swamp appear in Kaatz, "The Black Swamp" (page 145), and Winters's, *History of Northwest Ohio.* That people would try to spend the night leaning up against a tree is on p. 109 of Winters. That insects swarmed in thick clouds is found on page 162 of the William C. Holgate Journals, and reprinted in Louis Simonis, *Maumee River, 1835* (Defiance County Historical Society, 1979). That even Native tribes avoided the swamp comes from Frank N. Egerton, "Pollution and Aquatic Life in Lake Erie," (*Environmental Review,* 1987), page 197.

The word "fearsome" is used specifically to describe the swamp on page 126 of *A Memoir of Reverend Joseph Badger, Containing an Autobiography* (Sawyer, Ingersoll and Co., 1851). Bishop de Goesbriand's quote that "The Maumee Valley...devoured its inhabitants" is from his *Recollections,* also quoted in Nevin Winters, *A History of Northwest Ohio.*

p. 43: The Maumee and Western Reserve Road's reputation is recounted in Everett, *History of Sandusky County*, and repeated in Kaatz, "The Black Swamp." It's Kaatz who notes that Hull's Trace was even worse. That people would set up shop along the road to pull each other out is in C.E. Carter (ed.) *The Territory of Michigan, 1820-1829,* Vol. IX (*The Territorial Papers of the United States,* Washington: Gov't Printing Office, 1943).

p. 44: The story of how Antwerp got its name comes from "Name Calling: Origins of Paulding County's Place Names" by Melinda Krick (*Paulding Progress*, July 17, 2013).

p. 48: Information on the Wabash and Erie Canal came from sources already named, and from the website of the state of Indiana. For further reading, see "History of Canals in Indiana" by Howard Comstock in the *Indiana Quarterly Magazine of History*, 1911.

Workers on the Wabash and Erie Canal were paid in "Wild Cat" bills, a private paper currency borrowed out of Michigan. Until 1865, banks could issue banknotes without federal oversight. This was sometimes called "free banking." Supposedly, in Michigan, some banks were established way out where nobody could get to them—"where the wildcats roamed"—to prevent people from cashing in banknotes. Then the bankers would shut down the bank without ever having to pay anything out. The practice came to be known as "wildcat banking." There was soon a financial panic. Laborers on the canal went five months without being paid. Eventually this all got sorted out and everybody sat down to the work of canal-building. (See "Wildcat Banking, Banking Panics, and Free Banking in the United States" by

Gerald P. Dwyer, published in the December 1996 issue of *Economic Review*. Mention of canal workers being paid in "Wild Cat" bills, and that the bills were borrowed from Michigan, appears in Slocum, *History of the Maumee River Basin*, page 603.)

p. 49: In *A Hydrogeologic Atlas of Aquifers in Indiana* (1994), a Water-Resources Investigations Report for the USGS, Thomas Greeman notes the rapid rate at which trees were purged from the basin (page 8).

pp. 49-50: Cass and Woodbridge were quoted in Kaatz, "The Black Swamp," on pages. 8 and 10, respectively.

p. 50: John Johnson's drainage feat is in the Rochester *Democrat and Chronicle* (May 20, 2015), in an article titled "Finger Lakes Farmer's Invention Changed the World."

CHAPTER 4

p. 55-56: In 1787, the first two land ordinances were followed up by a third, officially titled, "An Ordinance for the Government of the Territory of the United States North-West of the River Ohio." It provided for a governor and secretary for the territory, and "magistrates and other civil officers" in counties and townships. A general assembly of citizens was to be established, which would regulate these leaders. The document also established trial-by-jury and *habeas corpus* in the Territory; set out a path to statehood for no more than five and no fewer than three states; encouraged morality, religion, and knowledge; and forbade slavery and involuntary servitude except in the conviction of crimes.

p. 56: I read about the Rectangular Survey System in *A History of the Rectangular Survey System* (1991) by C. Albert White, a report for the US Interior Bureau of Land Management, and in William Barrillas's *The Midwestern Pastoral* (Ohio University Press, 2006).

pp. 57-58: Details on Native removal come largely from *Contested Territories: Native Americans and Non-Natives in the Lower Great Lakes, 1700-1850,* by Charles Beatty-Medina and Melissa Rinehart (Lansing, MI: Michigan State University Press, 2002), especially the chapter "Miami Resistance and Resilience During the Removal Era" by Melissa Rinehart, and from the Miami Tribe of Oklahoma's website. The quotation about "handfuls of dirt" is found on page 152 of *Contested Territories*. Chief Richardville's claim that "we will not sell an inch of our land comes from Strack, et al., *myaamiaki aancihsaaciki: A Cultural Exploration of the Myaamia Removal Route*, 2009, page 2.

pp. 58-59: The problem of survey lines dividing an undividable landscape is outlined convincingly by Barrillas in the first chapter of *Midwestern Pastoral* (see above)—so much so that I borrowed his examples.

p. 59: Information on wetland and habitat loss comes from the US EPA Wetland Fact Sheet series, especially "Wetlands Overview" and "Functions and Values of Wetlands."

I borrow the term "biotic community" from Aldo Leopold, who used it in his book *A Sand County Almanac: And Sketches Here and There* (Oxford University Press, 1949).

pp. 59-60: Population numbers in the area formerly occupied by the Great Black Swamp come from Kaatz, "The Black Swamp: A

Study in Historical Geography" in *Annals of the Association of American Geographers,* Vol. 45, No. 1 (1955).

pp. 60-61: Information on agricultural development comes mainly from *Every Farm a Factory: The Industrial Ideal in American Agriculture* by Deborah Fitzgerald (Yale University Press, 2003); from a 2005 report for the US Dept. of Agriculture titled "The 20th Century Transformation of U.S. Agriculture and Farm Policy" (by C. Dimitri, A. Effland, and N. Conklin); from a 2002 article in the journal *Environmental Health Perspectives* titled "How Sustainable Agriculture can Address the Environmental and Human Health Harms of Industrial Agriculture"; and from "No-Till: The Quiet Revolution" by David Huggins and John Reganold in *Scientific American* (Vol. 288, No. 1, July 2008).

Information on tractor development comes from J.R. McNeill, *Something New Under the Sun: An Environmental History of the Twentieth-Century World* (W.W. Norton and Co., 2001), and from the February 2000 issue of *Illinois History.*

p. 66: The book I read in the Defiance County Library was *History of Defiance County, Ohio, Illustrated 1883* (Chicago: Warner, Beers & Co., 1883).

The idea of the triumph of civilization had staying power. In 1955, in a widely cited article on the Black Swamp, historical geographer Martin Kaatz wrote the following: "Within a few decades the Black Swamp was transformed from a useless, obstructive morass into one of the most productive regions in Ohio and the corn belt" (page 32).

CHAPTER 5

pp. 73-74: Sources of pollution are outlined in "Water-Resources Investigations Report 2000" a publication of the *United States Geological Survey* (D. Myers et al., 2000), and in other resources already cited.

Information on erosion—sheet, rill, and gully—and topsoil loss comes from the documentary *A Watershed Mentality* (2007), produced by Ray Steup for WFWA PBS39 in Fort Wayne; from Pimental, David and Donald Sparks, "Soil as an Endangered Ecosystem" in *BioScience* (Vol. 50, No. 11, 2000, page 947) and from Montgomery, David, "Soil Erosion and Agricultural Sustainability" in *Proceedings of the National Academy of Sciences of the United States of America* (Vol. 104, No. 33, August 14, 2007, pages 13268-13272).

Historical background on erosion comes largely from McNeill, *Something New Under the Sun: An Environmental History of the Twentieth-Century World*, especially the section titled "Three Pulses of Soil Erosion," and footnote 28 in the same chapter. The quote from Jacks and Whyte is there too.

p. 74: An Illinois blacksmith named John Deere invented the smooth steel "moldboard" plow in 1837. "Moldboard" refers to a particular design for the part of the plow that lifts and turns sod. Deere's invention was one of the primary tools in busting up the tallgrass prairies of the Midwest, and precipitated several environmental issues, including the Dust Bowl of the 1930s. (See "No-Till: The Quiet Revolution" by David Huggins and John Reganold in *Scientific American*, July 2008.)

pp. 75-76: Jeff Tyson of the Ohio Dimension of Wildlife, Noel Bulkhead of the US Geological Survey, and Todd Crail of the University of Toledo Department of Environmental Sciences all have pointed to soil runoff, sewage, and industrial pollutants, as well as phosphorous pollution and the loss of forest and wetlands, as factors in the decline of fish populations in the river. See "Toledo Magazine: A Changing Maumee River No Longer Home for Some Species" (July 6, 2014) by Matt Markey and Jeff Bastings, and "Seeking the Maumee River's Top Cat" (June 19, 2015) by Markey, both in the *Toledo Blade.* The latter is where I found the statistic on Toledo's record flathead catch, as well as just about anything else I might want to know about fishing for flathead.

Recommendations for fish consumption come from the July 2016 report, "Ohio Sport Fish Consumption Advisory," by the Ohio EPA Division of Surface Water.

p. 76: Information on PCBs comes from a US Environmental Protection Agency fact sheet on Polychlorinated Biphenyls, retrievable on the agency's website, and various other sources.

p. 76: The story about fingernails turning yellow comes from an April 5, 2015, story by Jamie Duffy in the *Journal Gazette,* titled "Getting to the Bottom of the Rivers." This article also provides an excellent overview of sources of pollution in and around Fort Wayne.

p. 78: In Fort Wayne, the city has used a combination of plants and limestone boulders to create a "riparian buffer" and keep soil, nutrients, and pesticides out of the water, especially along the St. Marys River. I read about this in "Fort Wayne Riverfront

Project Aims to Prevent Erosion, Pollution" by Lisa Ryan for Northeast Indiana Public Radio (WBOI). According to the article, "erosion washes away about a foot of soil each year on the banks of the St. Marys River."

p. 78: In 1943, Edward Faulkner, an agronomist from Elyria, Ohio, who had worked as an agricultural agent in the Erie basin, wrote a book called *The Ploughman's Folly. Nature* magazine called the book "an agricultural bombshell." In it, Faulkner challenged the received wisdom of the plow, given its effects on soil health and fertility. Faulkner was a conservationist at a time when conservation was lacking, and he was considered quite mad by many people. Eventually they came around. Research on modern no-till methods was pursued seriously beginning in the 1960s. (This is found, along with other information on soil erosion and no-till farming, in Huggins and Reganold, "No-Till"; additional details are from Faulkner's author page at Island Press.)

My definition of conservation tillage comes from page 73 of Huggins and Reganold, "No-Till." There are other aspects to it, but this is the gist as far as the Maumee River is concerned. Conservation tillage is also discussed at length in *A Watershed Mentality* (see above), and in "No-Till Farming is On the Rise. That's Actually a Big Deal" by Brad Plumer for the *Washington Post* (November 9, 2013).

p. 82: Father Joseph Pierre de Bonnecamps's journals are translated and published in *Ohio Archaeological and Historical Quarterly,*vol. XXIX, pages 397-423. The quote in this chapter comes from pages 411-412.

pp. 83-84: I read about the National Wild and Scenic Rivers Act of 1968 in a summary of the document available online at rivers.gov/wsr-act, and in a copy of the Act obtained through the Federal Energy Regulatory Commission. Information on the various designations of the Maumee River came from the Ohio Department of Natural Resources website.

p. 86: The verse quoted at the end of this chapter comes from Genesis 1:1-2 (New Revised Standard Version). Alternate translations for verse one include "When God began to create" or "In the beginning God created." Verse two could also be translated "the spirit of God" or "while a mighty wind."

CHAPTER 6

p. 90: I read about "buffalo rock" on an informational board in the park, and in an article titled "Roche de Boeuf to celebrate 'Buffalo Rock'" (*Toledo Blade*, September 22, 2010).

pp. 90-94: Most of the information on the Battle of Fallen Timbers and surrounding events comes from Alan Gaff, *Bayonets in the Wilderness: Anthony Wayne's Legion in the Old Northwest* (University of Oklahoma Press, 2004). Other information comes from *Tecumseh: A Life* by John Sugden (Henry Holt and Company, 1997). The quote about the Maumee "bubbling over a rocky sheet" comes from *Bayonets in the Wilderness*, page 301. The claim that the Wyandotte battle cry sounded like bells is on page 308. That howitzers made a sound akin to "Waugh! Waugh!" is detailed on the same page.

For a long time, people thought the Battle of Fallen Timbers took place where the limestone monument featuring Anthony Wayne is now. That's why the monument was put there in the first place. But in the 1980s, an archaeologist named G. Michael Pratt went back over the sites and concluded that the battle took place above the bluff a quarter mile downstream. Items found at the site included 524 spent lead shot (mostly buckshot); 11 lead shot with teeth marks—probably carried in the mouth for quick re-loading, and then spit out or lost during the battle; 111 buttons, 81 with a Federal insignia; a French socket bayonet; a gunflint and assorted lead gunflint wraps; the head of a small hatchet; and "a 1773 Spanish 2 Real coin exhibiting a suspension hole over the head of King Carlos." (see Pratt, "Remote Sensing Surveys at the Fallen Timbers Battlefield National Historic Site" in *Ohio Valley Historical Archaeology* [Vol. 18, 2003]).

In *Tecumseh: A Life,* John Sugden wrote that Wayne commanded 3,500 men, while the confederacy had only five hundred (page 89). Other sources stated that the confederacy had somewhere around eight or nine hundred soldiers. Wayne claimed that the number of Natives faced—and killed—was much higher than either of these estimates.

p. 93: Anthony Wayne is generally remembered as a brilliant leader and military genius. However, at least two eyewitness accounts of the Battle of Fallen Timbers are highly critical of Wayne's handling of the attack: *William Clark's Journal of General Wayne's Campaign* (compiled by R.C. McGrane, in the *Mississippi Valley Historical Review,* Vol. 1, No. 3, Dec. 1914), especially page 430; and *General James Wilkinson's Narrative of the Fallen Timbers Campaign* (compiled by M. M. Quaife for the *Mississippi Valley Historical Review,* Vol. 16, No. 1, June 1929). Wilkinson, who excoriates Wayne constantly in his account of the Fallen Timbers campaign, wrote, "…I pledge my Honor that I did not receive an order from him during the action, and Co[lonel] Hamtrack who command[ed] the left wing, assures me that his own case was exactly similar. Indeed I do no[t] hear of a single order which he gave except to poor M. Campble, who was sacrificed by a premature charge" (page 85). Later, he added, "It would appear from [General Wayne's] Letters & orders that his Bosom teemed with the finest sensations of humanity and philanthropy, and yet strange to tell, so little regard was paid to those who fell or Bled in the skirmish, that the Bodies of the former were left to ferment upon the surface, the prey of Vultures, untill the 22nd at which time & the Evening before unfortunate wounded men were found upon the field, where they had lain in agony from 10 oClock of the 20th" (page 86).

Wayne himself contradicted these allegations, writing that several of his commanders "rendered the most essential service by communicating my orders in every direction" (See Gaff, *Bayonets in the Wilderness*, page 306).

Both Wayne and Wilkinson's post-fighting assessments of the Battle of Fallen Timers come from *Bayonets in the Wilderness*, page 312.

pp. 93-94: Two years prior to the Battle of Fallen Timbers, in a letter to then-General Henry Knox, Wayne had suggested a modification to the design of the French musket that would allow a soldier to load on the run. This upset the standard pace of battles at the time, which were constrained by the needs of each party to stop and reload after every shot. At Fallen Timbers, it proved its worth. A captured Frenchman reported that the Natives' poor showing had to do with the rapid advance of the US Legion, who "would give them no time to load their pieces, but kept them constantly on the run."

It is said that Wayne told William Henry Harrison, his aide at the Battle of Fallen Timbers, "The standing order for the day is, 'Charge the damned rascals with the bayonet'" (See Gaff, *Bayonets in the Wilderness,* page 306).

p. 94: I learned about Tecumseh all over the place, but especially from *Tecumseh: A Life* by John Sugden, and the *Encyclopedia Brittanica* entry for "Tecumseh."

When the US failed to honor the Treaty of Greenville, Tecumseh was angry. Speaking of that treaty to William Henry Harrison in 1810, he said, "Since that treaty, here is how the Americans have treated us well: they have killed many Shawnee, many Winnebagos, many Miamis, many Delawares, and have taken land from them. When they killed them, no American ever was punished."

p. 95: The Ottawa River that runs through the University of Toledo campus is not the Ottawa River mentioned in the opening chapter of this book. That river, though it is also in Ohio, runs through Allen, Hardin, and Putnam counties, and drains into the Maumee from the south via the Auglaize River.

pp. 96-98: For information on algae blooms and cyanobacteria, I drew primarily from four sources: *Great Lakes: Natural History of a Changing Region* by Wayne Grady; *Streams: Their Ecology and Life* (2001) by Colbert Cushing and J. David Allan; a species outline for *Microsystis* put together by the Washington State Department of Ecology; various EPA reports, and a document published by the National Oceanic and Atmospheric Administration, titled "Nutrient Pollution—Eutrophication." That algae are sometimes called the "grasses of the waters" comes from Cushing and Allan, in *Streams,* page 153. Information regarding cyanobacteria reclassification is on page 156 of the same.

EPA warnings about cyanobacteria exposure come from a USEPA Region 9 report, "Frequently Asked Questions and Resources for Harmful Algal Blooms and Cyanobacterial Toxins" (Version 1, July 2015). Information on its carcinogenic potential come from the Washington State Department of Ecology.

The EPA is one major institution that continues to refer to cyanobacteria as algae.

p. 98: For a scientific discussion of the algae problem in Lake Erie and the Maumee River, see A. Michalak, E. Anderson, et al., "Record-Setting Algal Bloom in Lake Erie Caused by Agricultural and Meteorological Trends Consistent with Expected Future Conditions," in *Proceedings of the National Academy of Sciences* (of April 16, 2013); L. Backer and D. McGillicuddy, "Harmful Algae Blooms," in *Oceanography* (June, 2006); and David Biello, "Deadly Algae are Everywhere, Thanks to Agriculture," in *Scientific American* (August 8, 2014). For a good general read see *Erie: The Lake That Survived* by Noel Burns (1985).

pp. 98-99: The *Toledo Blade* actually ran two photographs of algae blooms dipped out of the city's water intake crib. The first appeared on August 3, 2014, as part of a series of photos by photographer Dave Zapotosky, titled "Algae at Toledo Water Intake." The second photo, also by Zapotosky, appeared the following day, accompanying a story by Tom Henry creatively titled "Algae's lake effect reveals pea green disaster."

A story published by the *Blade* on August 3, 2014, ran with the headline: "Water Crisis Grips Hundreds of Thousands in Toledo Area, State of Emergency Declared." The other articles mentioned in this paragraph are "2015 Lake Erie algae bloom largest on record," *Detroit Free Press* (November 11, 2015); and "Driven by Climate Change, Algae Blooms Behind Ohio Water Scare Are New Normal," by Jane Lee in *National Geographic* (August 6, 2014).

p. 99: This account of dwindling mayfly populations relies heavily on descriptions of mayfly habitation and migration patterns in Wayne Grady's *Great Lakes*, especially pages 234 and 236.

p. 99: Reports of Lake Erie's impending death (during the middle of the twentieth century) are recounted in Grady, *Great Lakes*.

p. 100: Frank Egerton's 1987 article in *Environmental Review*, "Pollution and Aquatic Life in Lake Erie: Early Scientific Studies," was helpful in gaining a sense of environmental awareness during the late nineteenth and early twentieth centuries. Especially pages 191 and 193.

pp. 100-101: For information on life along Lake Erie in the 1970s, I turned to Jerry Dennis's *The Living Great Lakes: Searching for the Heart of the Inland Seas* (Thomas Dunne Books, 2003). The claim that people largely believed the lake could handle their waste dumping appears on page 159.

The notion that people didn't think much about environmental issues before the 1970s comes from Egerton, "Pollution and Aquatic Life in Lake Erie: Early Scientific Studies," page 199.

By 1971, Lake Erie's reputation had sunk so low that even Dr. Seuss got in on the discussion with *The Lorax*:

You're glumping the pond where the Humming-Fish hummed!
No more can they hum, for their gills are all gummed.
So I'm sending them off. Oh, their future is dreary.
They'll walk on their fins and get woefully weary
In search of some water that isn't so smeary.
I hear things are just as bad up in Lake Erie.

p. 102: The 1975 report cited on Toledo pollution is "Water Pollution Investigation: Maumee River and Toledo Area," report by J. Horowitz of Enviro-Control, Inc., commissioned by the United States Environmental Protection Agency, January 1975, page 2.

p. 102: This chapter's definition of the "Rust Belt" comes from the editors of *Belt Magazine*, from the essay "Where is the Rust Belt?" in the edited volume *Dispatches from the Rust Belt* (2014), page 14.

p. 102: Toledo, Spain, is known for "Toledo steel"—a kind of steel that has a reputation for being particularly solid. This made it an excellent metal for use in sword blades, for which it became well known. Toledo, Spain, is believed by some to have been the namesake of Toledo, Ohio.

p. 103: Information on Toledo's industrial history comes largely from two short books: *Toledo: The 20th Century* (2005); and *Toledo: The 19th Century* (2004)—both by Barbara L. Floyd for Arcadia Publishing—and two articles in the *Toledo Blade*: "A Tradition of Industrial Growth" (March 3, 2003); and "Ex-Toledo Smelter Re-cast as Bridge Unit Fabricator" (May 5, 2003). That the

Veterans Glass City Skyway was assembled on the former site of a smelting factory is mentioned in the last of these.

Information on Toledo railroads comes from the website of the Toledo Port Authority, and from a map of the Toledo Terminal Railroad circa 1916. Information on the port of Toledo comes from the Toledo Port Authority, and an article in the *Toledo Blade* of January 18, 2016, titled "Shipping Tonnage Fell 30% in 2015," written by David Patch.

p. 104: Jesup W. Scott believed deeply in Toledo's future. "It is," he claimed, "the prairies, *only*, which has given Chicago the preference…the forest that has retarded the growth of Toledo." Now that the dense forests of the Great Black Swamp had been cleared, Scott reasoned, "the natural advantages" of Toledo's placement on Lake Erie and the Maumee River would award it superiority (*Toledo: Future Great City of the World*, 1868, page 40). Quoted material in this chapter comes from pages 28, 29, and 41.

p. 105: For more on Robert Howarth and Roxane Marino's work regarding nitrogen's effects on coastal waters, see "Nitrogen as the Limiting Nutrient for Eutrophication in Coastal Marine Ecosystems: Evolving Views over Three Decades," by Howarth and Marino in *Limnology & Oceanography* (Vol. 51, No. 1, 2006); and "Coastal Nitrogen Pollution: A Review of Sources and Trends Globally and Regionally" in *Harmful Algae* (Vol. 8, 2008).

p. 105: Information on the Toledo Water Reclamation Plant, and recent developments, comes from the city government's website, and from the website of the Toledo Waterways Initiative.

p. 105: Fort Wayne's wastewater mandate was reported by Randy

Spieth in "Fort Wayne Underground Pt. 1," for WANE-TV (June 24, 2014).

According to the website of Fort Wayne City Utilities, about one-third of Fort Wayne's sewer system is still combined-overflow.

pp. 105-106: For more on the Deep Rock Tunnel, see "Tunnel Works: Program Facts," a fact sheet produced by Fort Wayne City Utilities. Also see two articles by Kaitor Kpowsa for WANE-TV: "Fort Wayne's Biggest Public Works Project is Unveiled to Public" (February 25, 2016) and "Bids Open for Fort Wayne's $200M Sewer Overflow Tunnel" (February 16, 2017). The lowest initial bid was around $190 million dollars.

CHAPTER 7

pp. 109-111: Most of the information on the Toledo War and surrounding events comes from Don Faber's fascinating and well-written account in *The Toledo War*, published in 2010 by the University of Michigan Press, and from *The Facts and Historical Events of the Toledo War of 1835* by Willard Way (1859). Information on the Northern Ohio Boundary bill comes from *Michigan: A History of Explorers, Entrepreneurs, and Everyday People*, by Roger L. Rosentrater, (University of Michigan Press, 2013), page 99. Another important source was Frank Robson's "The Michigan and Ohio Boundary Line." *Michigan Historical Collections*, Vol. 11, 1905. The quote on Michigan's "Act to prevent the organization of foreign jurisdiction…" comes from page 222 of that document. Tana Mosier Porter's description of Benjamin Franklin Stickney is in *Toledo Profile: A Sesquicentennial History* (Toledo Sesquicentennial Commission, 1987). I first encountered it quoted in *The Toledo War*.

Frank Robson wrote in "The Michigan and Ohio Boundary Line" that "it is probable that...the importance of [Maumee Bay] had much to do in influencing the action of the Ohio authorities" (page 219). I believe him, especially considering that Ohio wanted to build a canal that would reach Lake Erie through the bay.

Regarding the Toledo Strip's uncertain status during the 1820s and 1830s—Letters have been found from this period addressed to "Postmaster, Tremainsville, State of Uncertainty." Tremainsville was a town in the Toledo Strip. (See *The Toledo War*, page 175.)

p. 111: "Armed" is a strong word to describe the men who accompanied Governor Mason to Toledo. A Michigander who went on this excursion later noted that while some of the men had guns, others were carrying only broom handles (see *The Toledo War*, page 162).

pp. 113-114: I read about silt in Maumee Bay, and the various efforts of people to figure out what to do with it, mainly in articles from the *Toledo Blade*. Two of the most useful were "$2.5 Million Facility to Store Silt Has Opened for Business" by Ignazio Messina (August 16, 2016), and "Water-Quality Law Dredges Up Unrest" by Tom Troy (May 4, 2015). I also learned a lot from the Toledo-Lucas County Port Authority's report on the Great Lakes Dredged Material Center for Innovation, prepared by Hull and Associates (May 21, 2015).

The quote from Craig Butler appears in "$2.5 Million Facility to Store Silt Has Opened for Business" (see above).

p. 115: Stories of successful wetland projects are in "Large-Scale Coastal Wetland Restoration on the Laurentian Great Lakes," by William Mitsch and Naiming Wang (*Ecological Engineering* July 2000), and in "A Natural Solution to Algae Problems in Western Lake Erie," Ray Stewart (*National Wetlands Newsletter*, May-June 2015), and elsewhere.

ACKNOWLEDGEMENTS

The project that would become this book began as an essay titled "Seeing the Maumee River," written for a course in nonfiction at the University of Illinois at Chicago and subsequently published in *Old Northwest Review*. But in truth it began much longer ago than that—perhaps as far back as childhood, when my siblings and I would go out with our "Nature Notebooks" and draw plants and animals in woods and by a creek near our house. Here are some of the people who have helped it along the way:

My parents and first teachers, Larry and Lora Schnurr, introduced me to the world and encouraged me to be curious. Their support in the years since has been unstinting. My siblings, Katie, Andrew, and Megan, in-laws Tom and Sue, and extended siblings Josh, Kenny, Ashley, Becca, and Jacob, have listened to my ramblings and sometimes even encouraged them.

Luis Urrea read an early draft of the essay "Seeing the Maumee River" and said, "I hate to break it to you, but this is a book." Then he taught me how to write one. (Luis, if you're reading—this whole thing is your fault.)

Many friends have tolerated me waxing on about the Maumee River over the past couple of years. Among those who have received the brunt of it are Brianne and Evan Cline, Jason and Emily Bleijerveld, Dirk Walker, Mark Rinehart, and the team at Anchor Films. The Bleijervelds also loaned me a lot of useful gear. Evan Cline is a trove of knowledge about the natural world, and helped me choose the right guidebooks.

Chris and Lisa Roseman, Kimberly and David Schneider, and Diana and Paul Bauer hosted me on my trip—sometimes with little to no notice—and made a tired traveler feel welcome. Aaron Schneider introduced me to his life and community; I cannot thank him enough.

I completed much of the research for this project while on a fellowship in American Studies at Purdue University. Dr. Rayvon Fouché gave thoughtful feedback on an early draft of the manuscript. Lettie Haver of ACRES Land Trust shared additional knowledge and research materials with me. I am grateful to ACRES Executive Director Jason Kissel for granting me permission to camp in a preserve.

Jason Bleijerveld, in addition to accompanying me for some of the more grueling sections of the trip, read and commented on early chapter drafts. His notes were invaluable. Jason also contributed one of my favorite descriptions in the entire book—that our canoe behaved "like a large, unruly dog."

Speaking of dogs—Thor, a devoted Pomeranian-Chihuahua mixed breed, was my constant office companion during long reading and writing sessions.

Anne Trubek took a chance on an unknown writer and provided valuable guidance and insight. Anne is perhaps the only person who shares my enthusiasm for the Great Black Swamp. She and

Martha Bayne worked over the manuscript in detail; the book is better for them.

Anna Schnurr has lived with this book as long as I have. It is not easy to be partnered with a writer, and she weathers the task with grace. Her contribution to *In the Watershed* is impossible to summarize; this book belongs as much to her as it does me.

Any aspect of this book that is good, that is thoughtful, that rings of truth and beauty, is due in large part to the influences of those named above. Its faults, and any sins of error or omission, are mine alone.

ABOUT THE AUTHOR

Ryan Schnurr is a writer and photographer from northeast Indiana. His work has appeared or is forthcoming in two anthologies of Midwestern writing, and has been published by *Midwestern Gothic*, *Old Northwest Review*, and *Belt Magazine*, among others. *In the Watershed* is his first book.